COMET HALLEY
Once in a Lifetime

Comet Halley on June 6, 1910 (photograph by E. E. Barnard courtesy of Yerkes Observatory)

COMET HALLEY
Once in a Lifetime

MARK LITTMANN

DONALD K. YEOMANS

AMERICAN CHEMICAL SOCIETY
Washington, D.C.
1985

Library of Congress Cataloging in Publication Data

Littmann, Mark, 1939–
  Comet Halley: once in a lifetime.

  Bibliography: p.
  Includes index.

  1. Halley's comet.

  I. Yeomans, Donald K. II. Title

QB723.H2L56   1985   523.6′4   85–4012
ISBN 0–8412–0905–7
ISBN 0–8412–0911–1 (pbk.)

MARK LITTMANN

He is the Science Communicator for NASA's Space Telescope Science Institute at Johns Hopkins University in Baltimore. Previously, he was Director of the Hansen Planetarium in Salt Lake City for eighteen years where he pioneered many innovations in planetarium programming and received two special awards from his colleagues for distinguished service to the planetarium profession. His programs, including three under American Chemical Society sponsorship, are performed at planetariums throughout the world. In Salt Lake City, Dr. Littmann also taught astronomy and literature at the University of Utah and Westminster College.

Dr. Littmann received his B.S. in chemistry and literature from the Massachusetts Institute of Technology, his M.A. in creative writing from Hollins College, and his Ph.D. in English from Northwestern University.

Mark and his wife, Peggy, have a daughter, Beth.

DONALD K. YEOMANS

As an astronomer at the Jet Propulsion Laboratory, he provided the accurate position predictions that led to the successful recovery of Comet Halley at Mt. Palomar on October 16, 1982. Using ancient Chinese observations, he has traced the motion of Comet Halley back to 1404 B.C. During 1985–86, he will provide observers with the predictions necessary for studying this famous comet. In connection with the planned comet flyby missions in 1985–86, Dr. Yeomans has been working closely with the flight projects in the United States, Europe, Japan, and the Soviet Union. Since receiving his Ph.D. in astronomy from the University of Maryland, he has written over forty published works and is currently a member of several international groups of scientists planning future ground-based and space observations of comets.

Dr. Yeomans lives with his wife and two children in La Cañada, California.

DEDICATION

From Mark Littmann
to Lewis and Muriel Littmann
and Peggy

From Donald K. Yeomans
to Laurie, Sarah, and Keith

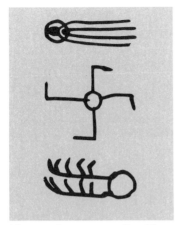

Three types of comets represented in a Chinese comet catalog of 168 B.C.

CONTENTS

COMING IN FROM
THE COLD

On October 16, 1982, at Palomar Observatory in California, the great 200-inch telescope peered into the darkness. Here, and at large observatories throughout the world, astronomers were searching for a celestial visitor approximately 3.7 miles (6 kilometers) in diameter and 1 billion miles away. The 200-inch mirror reflected starlight onto a charge-coupled device, a highly sensitive electronic eye that recorded light fifty million times fainter than the human eye can see.

A few days later, after many hours of processing and enhancing the images, David C. Jewitt and G. Edward Danielson detected a dot where no star existed, a dot almost exactly where Donald K. Yeomans had calculated the onrushing guest would be.[1]

The comet was coming back again, just as it had for more than 2200 years.

It is our turn now—Comet Halley—once in a lifetime.

1

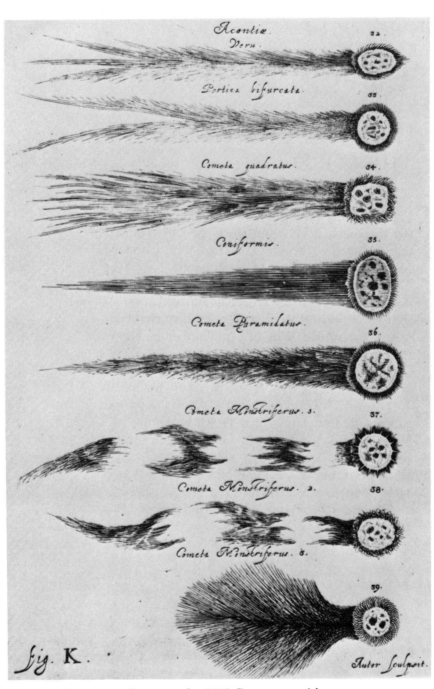

Types of cometary forms in the 1668 Cometographia
by Johannes Hevelius (from the collection of D. K. Yeomans)

Chapter One

A COMET
THROUGH THE AGES

When beggars die, there are no comets seen;
The heavens themselves blaze forth the death of princes.
—WILLIAM SHAKESPEARE
(*Julius Caesar*, Act II, Scene 2)

A HAIRY STAR

For centuries the sight of a comet, even the word "comet," struck fear to the hearts of people.

One particular comet appeared in the skies around the world in the year 240 B.C. For the Greeks, a comet was not a visitor from space. Most accepted the opinion of the philosopher Aristotle who had taught 100 years earlier that comets were part of the Earth's atmosphere.

Aristotle believed the universe was composed of two parts. Below was the Earth, the realm of change and imperfection. The region above, where the Moon, Sun, planets, and stars lay, was the celestial realm of perfection. Circles were the perfect shape, so everything in the celestial abode traveled around the Earth in circles, and in that realm nothing changed.

Comets, Aristotle observed, do not move in circles and are continuously changing. Therefore, he concluded, they must be close to Earth. He further explained that comets were produced when the hot dry vapors of our world rose to the top of our atmosphere

3

and were struck by a "fiery principle" from above. The appearance of comets was a sign of coming wind and drought.[2]

Aristotle's concept was the accepted view of the universe for the Western world during the next 1900 years.

The word "comet" also comes from the Greeks. Comet comes from *kometes*, meaning "hairy one"—a star with long hair.[3]

HARBINGERS OF WOE

Centuries passed, and with them came a succession of comets— unannounced, unpredictable objects of fear and superstition. No matter when a comet appeared, there was always a war, epidemic, famine, or natural disaster in the past, in progress, or soon to come for which the comet could be blamed.

THE COMET AND THE HUN

A comet was seen around the world in A.D. 451. As usual, it was interpreted as a bad omen, but for whom?

During the previous century, the Huns from central Asia had pushed westward until they dominated central Europe. Their newest leader was even more aggressive, defeating Roman armies, displacing people all across Europe, and forcing Rome to pay a heavy annual tribute in gold. This leader's name was Attila the Hun.

In A.D. 451, the Roman general Aetius persuaded the Visigoth leader Theodoric to take the field against Attila. Theodoric was killed, but his army defeated the Huns in the Battle of Châlons, the only actual defeat in war that Attila ever sustained. After the battle was over, the Romans decided that the comet had indeed been an evil omen—for Attila.

THE COMET AND THE KING

In A.D. 837, a most unusual comet appeared and kept brightening until it could have been briefly visible in daylight. Its tail stretched across more than half the sky.

In A.D. 837, Louis I, son of Charlemagne, ruled the Frankish and Roman Empires. He saw the comet as an omen "foreboding

DE OCCVLTATIONE ET APPARITIONE
Cometæ. Caput decimumicxsrtus.

AMETSI Cometa 33 ingreſſi die nobis obſpectus ſit primo, non deſiunt tamen quiruendum ſe vidiſſe ſexto ſeptimoque affirmare, vere autem pendus ab occidere paucis ab oriente viſum eſt, & perfecto fieri non poteſt quin ita fieri non ſit, fieri non omnibus apparuerit, non enim ita multo poſt ortum ſuum debuit videri mane deſit, ea de cauſa, quod facunde magis ac magis cõſinicum in certum feſtinant. Nam die 18, ut viderat in ſequentibus eſt, orti cum ipſo Sole cepit, quoniã a præpinquior erat tam Sol 33 de quaq; ut certi pouetit, ortum parens eim Sole. Inde factum eſt, utplerãq; imperitioribus alius ab nude cometa fuſſe putaretur, quaſi alio fuerit, vitus ab oriente, alter in occidente, neſcentibus tam in ortu quàm in occaſu apparere poſſe, non ſicut atq; ſtella alum. Id quod etiam Matthæo Palmerio Florentino impoſuit in cronicis ſuis aſſeribus binos in Ianuario cometa fuſiſſe, anno poſt Chriſti 712, altumq; Solem mixtiſt, alium vero ſeldiquorum, cum reuera non duo, ſed viuus & idem fuerit, nunc ante minus poſt Solem, ea quæ dixi rationi, bicens. Sed ad rem, Sol cum 32 Leonis gradum poſſideret, occidebat hora poſt meridiem 6 uni, 34, Cometa vero hora 8 m, 55, & declinatio eius aſcenſiu recta conſideretur. Vide ſuper Solem fuſſe ſub horizonte hora 30 m, 32, A Cometam horã 4 m, 16, cuius ideirco ortus hora 4 m, 30 poſt noctu medium eſſe debuit. Cum vero de ortu & occaſu cometæ loquor, non ſabuda velo, nihil viſum per inſtrumenta eo tempore quo horizontes contigerit Cometæ, illud enim ſtrinde declinatio aſcenſiuq; recta eiuſſemd cognita fuerit ſing, & ſi vero per inſtrumentum id fieri cõnatus, cenſeri tamen nequaq; potuiſſet, quoties enim horizonti propinquior fieret 3 vel 4 gra, interuallo, flamma eius extingui omnino videbatur, adeo, ut ſi quando à me ſub nubila condi crederes ſit, donec admoneret Plinius libro naturali hiſtoriæ 2 cap, 12, Cometas in occaſu raroq; poſſe nunquam eſſe, id eſt, nunquam apparere. In enim genuinus Plinii intellectuuif, ni fallor, quem eo libentius ego quoq; adorno, quoniam ita oculis eſſe comperiris. De viſ meteoroſcopi nihil amplius agi, ſed poſt finem cometæ, motum quoq; diutinum per meteoroſcopium motiuit docebo, ſufficiat autem hanc viam obſeruationis methodiica fuiſſe, & quaſi per membra tradidiſſe, in ſequentibus enim id prægir obſeruationes meras per meteoroſcopium cõel, quotas propoſitionã. Figuræ præterea ſequentis oſtendiunt altitudines Cometarum donec exitat, ſupra horizontem exiſtentium tempore quo Sol occumbit, illud ſue quaſ profundus Sol exiſtere ſub ſunitore coneta eunde Kandes.

ſuſq; qui contingunt 33 die Auguſti.

ſecunda Cometæ obſeruatio peracta eſt.

- Altitudo Cometæ ſupra horizontem, gra, 6 mi, 12
- Azimuth Cometæ ab occaſu Septen, verſus gr, 45 mi, 11
- Altitudo extremitatis caudæ gradibj 11 mi, 18
- Azimuth extremitatis caudinæ Septentrio, gra, 57 mi, 30

Hæc ſequentia ex obſeruatione conſurgunt.

Latitudo Cometæ gra, 33 & 1, Locus Cometæ gra, 2 mi, 50 &c.
Declinatio Cometæ, 37 m, 31 Sept. Aſcõſio recta Co. gr, 121 m r
Co. Meduat cõlõ ho, 13 ab, Amplitudo ot A occ, ea gr, 12 m Sep.
Occa. Com. hora 2 m 32 poſt m. Com. occidet cſ 11 gra, 16
Diſtantia Com. à Sole gr, 13 bi, 40, Ortus Com. horã 2 mi, 12
Loc'extrem caudæ gr, 320 30 & Latitu, extre gr 32 m 40 Sept.

Situs Cometæ occaſus tempore. — Situs Cometæ ortus tempore.

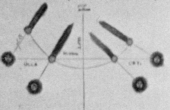

ſeruatio tertia facta eſt.

- Altitudo Cometæ ſupra horizontem gra, 9
- Azimuth Cometæ ab occ, verſus Sept, gra, 41 mi, 11,
- Altitudo extremitatis caudæ ſupra horizon, gra, 30 mi, 9
- Azimuth huius extremitatis gra, 50

Hæc autem obſeruatio collegit ea quæ ſequuntur.

Latitudo Come, gra, 11, Locus verus Come, gra, 14 mi, 20 33
Dech. Come. gr, 33 m 30 Sep. Aſcen. recta Co, 15 r gra, 21 mi.
Latitudo extremitatis caudæ ab ecliptica gra, 34 mi, 31
Locus verus in ecliptica extremæ caudæ, 23 gra, 14 mi, 37
Diſtan. à Sole gra, 13 mi, 33. Aſcend at. Come. hora 18 m, 24
Occa. Come. hõ, 2 m, 44 poſt. Or. Come. horã 2 mi, 30 ant.
Hac die Com. bcliace accidit, ut nõ amplius ante Solem orii. cerneret.

Situs Cometæ occaſus tempore. — Situs Cometæ ortus tempore.

conſideratio Cometæ quarta.

- Altitudo Cometæ ſupra hori. gra, 2 mi, 43
- Azimuth Cometæ ab occ.verſus Sept. gra, 37 mi, 33.
- Super caudã nihil ulterius agem' ſubobſcut enim prius, per quæ ſatis liquet à Sole caudã auerti, quoniã in poſteri de hac ſigle debimus.

Talia ex obſeruatione conſtant.

Latitudo Cometæ ab ecliptica gra, 12 mi, 3
Verus locus Cometæ in ecliptica gra, 6 mi, 33 ly.
Declinatio Cometæ ab æquinoc 33 gra, 33 mi. Septentrio.
Aſcenſio recta Cometæ 105 gra, 33 mi.

Peter Apian provided these illustrations of his discovery that comet tails flow away from the Sun. From Astronomicum Caesareum *(Leipzig, Edition Leipzig, 1967), a facsimile of the Ingolstadt 1540 edition.*

Roman silver coin (denarius) issued during the reign of the first Roman Emperor Augustus (27 B.C.–A.D. 14) with the head of Augustus Caesar on one side and a stylized representation of a comet on the reverse side. On the obverse side of the coin is the partially obliterated inscription "Caesar Augustus" and on the reverse side is "Divvus Julius" (Divine Julius) (from the collection of D. K. Yeomans).

THE COMETS OF 44 B.C. AND 12 B.C.

Julius Caesar was assassinated in the Roman Senate on March 15, 44 B.C. While Caesar's adopted son Octavius (later Emperor Augustus) was holding athletic games in honor of the fallen Caesar, a comet (not Halley's) appeared in the northern heavens for seven days. In the first century A.D., the Roman historian Pliny recorded Caesar's comet as a remarkable one seen by all the world. It rose after sundown, and the common people believed it signified the reception of the soul of Caesar into the realm of immortal gods.

With the death of Julius Caesar, a power struggle ensued between Octavius and Mark Antony. The struggle finally was resolved in favor of Octavius after the 31 B.C. Battle of Actium on the western coast of Greece in which Marcus Agrippa, Octavius' general, defeated the forces of Antony and Cleopatra in decisive naval battles. Octavius became Emperor Augustus.

When Agrippa died in 12 B.C., an impressive funeral celebration was staged in his honor by the grateful Emperor Augustus. Agrippa was also honored just prior to his death by the appearance of Comet Halley, which made a close approach to the Earth in September 12 B.C. and appeared as bright as the brighter stars. Some 200 years later, the Roman historian Dio Cassius wrote of Comet Halley as a sword suspended over Rome before the death of Agrippa.

and sad...'For they say that by this token a change in the realm and the death of a prince are made known.'"[4] Louis the Debonair, or Louis the Pious as he became better known, then stayed up all night as a vigil, praying to God, and when morning came he commanded that generous alms be given to the poor and that larger contributions be made to churches and monasteries.

He did not receive the expected benevolence from heaven. One of his sons, Pepin, died the next year, and Louis himself died three years later.

European observers through the Middle Ages usually recorded such comet appearances in tones of hysteria. Unknown to the Europeans, the Chinese also observed this comet. The Chinese had made a specialty of observing what they called "broom stars" throughout recorded history.[5] Unlike European observations, apparently written in panic, the Chinese observations carefully recorded comet positions, motions, and appearances. Their records were so good that certainly by A.D. 837, and most likely as early as A.D. 635, Chinese astronomers had reached the conclusion that comet tails extend away from the Sun. They also taught that comets shine by reflected sunlight, just as the Moon does.[6]

A comet's tail extends away from the Sun. A comet's tail lengthens as it approaches the Sun and shortens (more slowly) as the comet recedes.

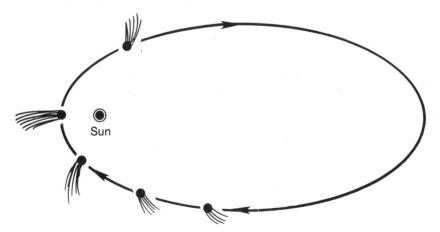

EARLY CHINESE COMETARY OBSERVATIONS

The Chinese were the most accurate and prolific astronomical observers in the world, unmatched until the rise of Arabic astronomy in the eleventh century. Through the tenth century, Chinese records were virtually the only significant collection of astronomical observations. Unlike the ancient Greeks, Chinese astronomers did not concern themselves with theory, such as developing geometrical formulations for the motions of the heavenly bodies. The Chinese were only interested in what they actually saw, and fortunately they recorded nearly everything that was out of the ordinary. Chinese observations of comets, as well as stars, novas, and planets, were required in prognosticating for state affairs. Unlike the Greek astronomers who were strictly academics, the Chinese astronomers were intimately connected to the ruling sovereigns and even resided within the walls of the imperial palace. They were continuously called upon to guide the emperor with astrological advice.

Chinese astronomy was well ahead of European astronomy up to and including the fourteenth century. Accurate and detailed Chinese star maps are extant from A.D. 940 through the fourteenth century, a period when Europe had nothing comparable. The Chinese continuously developed new and complex astronomical instruments. In 1090, they coupled an armillary shere (for measuring positions of objects in the sky) to a water-powered clock drive. During the thirteenth century, the Chinese invented a mounting for their observational devices (the equatorial mount) that was designed to follow the motion of celestial objects in the sky. However, the most important Chinese contribution to astronomy remains the dynastic records of continuous observations of celestial phenomena such as eclipses, novas, sunspots, and comets.

The first extant description of cometary forms was a Chinese catalog on silk recently found in the 168 B.C. Han Tomb in Hunan Province, China. These cometary forms from the catalog have been redrawn for clarity.

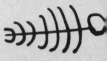

THE COMET AND THE CONQUEROR

A comet was visible in the spring of 1066. In Europe, on either side of the English Channel, two armies prepared for war. King Harold of England worried about threats of invasion by William, Duke of Normandy. William's armies crossed the English Channel and on October 14, 1066, defeated and killed King Harold in the Battle of Hastings. Duke William became known as William the Conqueror,[7] and the course of world history was radically changed by the Norman Conquest.

In honor of the victory, the artisans of Bayeux, in Normandy, created by embroidery an elaborate panorama of dramatic scenes, 231 feet long and 19½ inches high, now cherished as one of the art treasures of the world: the Bayeux Tapestry.[8]

The Comet of 1066 was so startling and impressive that it was given a prominent place on the Bayeux Tapestry, its glowing head pulling a train of fire. Beneath the comet, to one side, King Harold of England listens as an advisor warns him about the evil omen passing over his head. On the other side, the common people point to the comet and cringe with fear. The Latin inscription tells us, "They marvel at the star."

THE COMET AND THE ARTIST

A comet visible in 1301 so impressed an Italian artist named Giotto di Bondone that he incorporated it into one of the frescoes he was painting on the walls of the Scrovegni (Arena) Chapel in Padua. He finished "The Adoration of the Magi" about 1304, which showed the Wise Men from the East paying homage to the infant Jesus. Above the manger, Giotto depicted rather realistically a comet as the star of Bethlehem.

THE COMET IN PRAYER

A "terrible" comet "of extraordinary magnitude" appeared in 1456. Turkish forces under Mohammed II had seized Constantinople three years earlier, had advanced across Europe, and in 1456 were poised to attack Belgrade, which was defended by the armies of Pope Calixtus III. Throughout Europe, Catholics prayed as they thought the Pope had directed: "Dear God, save us from the Devil, the Turk, and the Comet."[9]

The sight of so bright a comet terrified and paralyzed both armies. Then János Hunyadi, the Catholic general, told his troops that the comet foretold the defeat of the Turks. The Papal army attacked, and the Turkish invasion was halted.

COMET HALLEY AND THE STAR OF BETHLEHEM

At one time or another the star of Bethlehem has been attributed to a miracle, comet, bright meteor, dense meteor shower, Venus, Mars, Jupiter, Saturn, nova or supernova, bright star, constellation, zodiacal light, or a combination of two or more of these. Many recent scholars attribute the star of Bethlehem to a triple conjunction of Saturn and Jupiter that occurred during 7 and 6 B.C. or to conjunctions of Jupiter and Venus in 3 and 2 B.C. The star of Bethlehem also may have been invented after the birth of Jesus by an overzealous Christian scholar.

An identification of the star of Bethlehem might be easier if we knew when Jesus was born, but that date is highly uncertain. According to the book of Luke, Jesus and John the Baptist differed in age by only six months, and John began his ministry in the fifteenth year of the reign of Tiberius (in A.D. 29). If John began his ministry when approximately thirty years of age, then John (and Jesus) would have been born in approximately 2 B.C. (there is no year zero). On the other hand, Jesus must have been born prior to the death of King Herod, who is thought to have died in 4 B.C. Mary and Joseph were making the long journey from Nazareth to Bethlehem to pay their tax, and, according to Luke, the taxation took place when Quirinius was governor of Syria. Quirinius was not appointed governor of Syria until A.D. 6. Hence, the biblical evidence for the date of Jesus' birth is quite contradictory.

THE COMET IN SCIENCE

By 1531, when a notable comet again steered a majestic path through the skies, a new age of European science was dawning. In 1532, Peter Apian in Germany published *Practica* which showed on its title page a woodcut of a comet with its tail extending away from the Sun, some 900 years after the Chinese made this

The only biblical reference to the star of Bethlehem is in the gospel of Matthew:

Now when Jesus was born in Bethlehem of Judea in the days of Herod the king, behold, there came wise men from the east to Jerusalem, saying, where is he that is born King of the Jews? For we have seen his star in the east, and are come to worship him.... when they had heard the king, they departed; and, lo, the star, which they saw in the east, went before them, till it came and stood over where the young child was.

Although 12 B.C. is almost certainly too early for the birth of Jesus and although comets were almost always considered to be bad omens, it is interesting that the 12 B.C. apparition of Comet Halley would fit the biblical description of the star of Bethlehem. The Chinese recorded the 12 B.C. apparition of Comet Halley as lasting fifty-six days. In early September of 12 B.C., an observer located in the Middle East would have seen Comet Halley as an obvious early morning object above the eastern horizon. Later in September, the comet would have appeared as a bright early evening object above the western horizon. Like the biblical star of Bethlehem, Comet Halley would have been first seen in the east and then later in the west, as if to guide the wise men westward to Jerusalem.

If a star of Bethlehem did exist, it was more likely a subtle grouping or motion of celestial bodies of interest to astrologers such as the Magi from Persia. The New Testament indicates that no one except the Magi noticed the "Christmas star."

The Comet of 1577 as depicted on this broadside printed by Peter Codicillus in Prague. An artist is seen drawing the comet and is aided by men holding his sketchbook and lantern. The heading reads, "concerning the fearful and wonderful comet that appeared in the sky on the Tuesday after Martinmass (November 12) of this year 1577." From the Wikiana collection at the Zentralbibliothek, Zurich, Switzerland (photograph courtesy of O. Gingerich).

discovery. Meanwhile, Girolamo Fracastoro in Italy independently made the same discovery, which he published in his *Homocentrica* in 1538. In 1540, Apian elaborated on his own discovery in his beautifully illustrated *Astronomicum Caesareum* in which he showed in color and explained in detail that comet tails flow away from the Sun.[10]

In Poland during this period, a mathematician named Nicolaus Copernicus was quietly circulating among scientific friends a manuscript in which he suggested that the planets traveled around the Sun, not the Earth.

A BREAKTHROUGH

In 1577, a major comet burst into visibility over Europe. A new generation of scientists was at work, including a Danish astronomer named Tycho Brahe.

When he was nineteen years old, Tycho quarreled with another student about a mathematics problem. They settled the argument by fighting a duel, and in that swordfight Tycho had his nose cut off. He replaced it with one made of gold and silver, and he always kept it well polished.[11]

Tycho was eccentric, but he was the finest observer of his age. He was so skillful that the King of Denmark gave him the island of Hven and all the money he needed to build the best observatory in the world before the invention of the telescope. When the comet appeared, Tycho was ready.

The people of 1577 still thought that comets were what Aristotle had described nineteen centuries earlier: flaming vapors

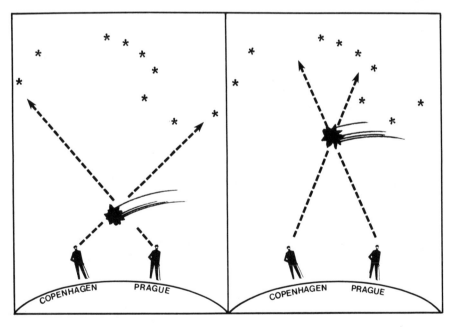

Tycho Brahe used parallax measurements to demonstrate that the Comet of 1577 lay far beyond the Moon, contrary to the belief of the day. Tycho observed the comet from Denmark and a colleague observed it from Prague. If the comet were close to Earth, observers in different locations would see the comet against a substantially different stellar background. The smaller the shift against the stellar background, the farther away the object must be.

*Johannes Kepler
(photograph courtesy of
Yerkes Observatory)*

high in the Earth's atmosphere. Christianity not only adopted this concept but added a twist of its own. Comets were the sins of mankind rising skyward as noxious gases. God did not want his heavens polluted by the sins of man, so He smote them in his fury, causing them to blaze forth as warnings to the sinners of Earth.

In response to the Comet of 1577, a Lutheran bishop named Andreas Celichius published a book in 1578 entitled (in English) *Theological Reminder of the New Comet*. Comets, he said, are

> the thick smoke of human sins, rising every day, every hour, every moment full of stench and horror, before the face of God, and becoming gradually so thick as to form a comet, with curled and plaited tresses, which at last is kindled by the hot and fiery anger of the Supreme Heavenly Judge.[12]

Such was the prevailing doctrine. But observations and measurements meant far more than doctrine to Tycho Brahe. If comets were flaming vapors in the upper atmosphere of Earth only a few miles high, then Tycho in Denmark would see the comet against one background of stars and an observer in Prague would see that same comet at the same time against a slightly different background. This shift is called parallax. The smaller the parallax, the farther away an object is.

Tycho observed the comet carefully and collected reports from other astronomers. The comet showed too little parallax to be in the atmosphere of Earth. In fact, the comet showed less parallax than the Moon. It had to be at least four times farther away.[13]

Yet Aristotle had taught that the realm of the Moon, Sun, and planets was perfect and unchanging. Tycho proved Aristotle was wrong. A comet was wandering among the planets and changing as it went.[14] The old ideas began to crumble.

THE COMET IN A NEW CENTURY

Another comet loomed into view in 1607. Tycho had died and left his lifetime of planet observations to his assistant, Johannes Kepler. Kepler, a fine mathematician, was in the process of using those observations to demonstrate that the planets actually do travel

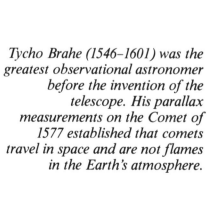

Tycho Brahe (1546–1601) was the greatest observational astronomer before the invention of the telescope. His parallax measurements on the Comet of 1577 established that comets travel in space and are not flames in the Earth's atmosphere.

around the Sun, but not in circles as Copernicus thought. The planets orbit the Sun in ellipses. Although he could not explain this phenomenon, Kepler demonstrated that as each planet travels along its elliptical orbit, it moves faster when it is close to the Sun and slower when it is farther away.

Kepler also was the first to suggest that comet tails are created by the pressure of sunlight, that comets will eventually disappear because of the material they shed, and that comets are "as numerous in the heavens as are the fish in the sea."[15]

Still, for most people as the seventeenth century began, the nature of a comet remained obscured by superstition. Shakespeare's plays were not nearly as popular through that period as a continuously translated and reprinted book by Guillaume du Bartas which had this passage to offer:

> There, with long bloody hair, a blazing star
> Threatens the world with famine, plague, and war;
> To princes, death; to kingdoms many crosses;
> To all estates, inevitable losses;
> To herd[s]men, rot; to ploughmen, hapless seasons;
> To sailors, storms; to cities, civil treasons.[16]

Chapter Two

THE MAN
WITH THE ANSWER

I may, therefore, with some confidence,
predict its return in the year 1758.
—EDMOND HALLEY

A MAN NAMED HALLEY

In 1682, a fine comet appeared in the heavens. A twenty-six-year-old Englishman plotted the changing positions of the comet very carefully. By age twenty-two, he had already spent a year south of the equator to produce the first star catalogue of the southern skies. Before his long and distinguished scientific career was over, Edmond Halley would lay the foundations of physical oceanography, geomagnetism, and life insurance statistics; furthermore, he would make major contributions to astronomy, meteorology, navigation, mathematics, and optics. He was also the first to invent practical deep-sea-diving equipment. But most people today know of him only because a special comet bears his name.

In early 1684, twenty-eight-year-old Edmond Halley, architect Christopher Wren, and physicist Robert Hooke—three of the most distinguished scientists in the world at that time—were meeting regularly at Royal Society gatherings and at London coffeehouses to discuss the great mysteries of science.[1] In particular, they discussed a force that seemed to hold the Moon in orbit around the

17

Edmond Halley as painted by Thomas Murray in 1686 when Halley was thirty years old. The titles were written on later (reproduced with permission from the Royal Society).

ABOUT HALLEY'S NAME

How did Halley pronounce his name? Because English spelling in Halley's day was not yet standardized, we may gain some indication of the pronunciation by the way people spelled Halley's name trying to get the right sound phonetically. His name was sometimes written as Hawley and Hall-ey, but Haley and other spellings are also found.

Several writers carried out an interesting experiment. They opened the present-day London telephone directory, called people with the last name Halley, and asked them how they pronounced their name. Almost all pronounced their name "HAL-ey," so it is likely that Halley pronounced his name so that it rhymed with alley. This and similar letter combinations appear frequently in English with this same pronunciation: galley, valley, dally, rally, tally, Sally, and O'Malley. The pronunciation "HAIL-ey" is almost surely wrong.

Halley's first name was Edmond, although many books spell it Edmund. We know Halley's spelling of his name because he signed his name Edmond on his will and other crucial documents. The misspelling Edmund may arise in part from a portrait of Halley with this caption across the top: "Edmund. Halleivs LL.D. Geom. Prof. Savil. & R.S. Secret." This caption, inscribed at a later date, uses the Latinized spelling and abbreviations of Halley's name and (future) positions: Edmundus Halleius, Doctor of Laws (honorary degree), Savilian Professor of Geometry (at Oxford) and Secretary of the Royal Society.

Earth and the planets in orbit around the Sun. But how did this force operate, and how could it be demonstrated?

None of them could produce an answer. Wren said he had tried, but the mathematics had defeated him. Hooke said he already had the answer, but was withholding it so that everyone would be properly appreciative when he stepped forth.[2] Wren offered a prize, a very expensive book of their choice, if either Hooke or Halley could submit a proof. Time passed, and nothing was forthcoming from Hooke.

But all three men knew of a professor at Cambridge University who seldom published anything, yet was good at math and might have an idea about the Moon and planets problem. Halley decided to visit this mathematician. His name was Isaac Newton.

A MEETING WITH NEWTON

Newton was born fourteen years before Halley, and his family had expected him to become a farmer. Newton showed no talent for

Isaac Newton (photograph courtesy of Yerkes Observatory)

THE COMET EGG OF 1680

The Great Comet of 1680 was discovered on November 14, 1680, by the German almanac maker Gottfried Kirch. As the father of fourteen children, one wonders how Kirch found the time for his systematic searches of the night skies. Kirch was the very first to discover a comet with the help of a telescope, and he published his observations of the impressive comet the following year. He believed that although comets may be natural phenomena, they are nonetheless signs from God. Kirch's moderate comments on this comet were a far cry from the majority view. In Rome even the fowls of the barnyard responded to the comet's appearance, and a letter to the prestigious French Academy of Sciences announced that a Roman hen had laid an egg on which was shown the comet. The letter reported that on the night of December 2, 1680, the hen, which had never before laid an egg, began to cluck in a loud and extraordinary fashion. She succeeded in laying an enormous egg that had natural markings resembling a comet on a stellar background. A representative of the French Academy, after apologizing for taking notice of the occurrence, explained that the egg was not marked with a comet as many had believed, but rather it was marked with several stars. A drawing was included in his report to prove the point. Nevertheless, some important people in Rome vouched for the comet egg in letters to their colleagues in France and Germany. Several pictures of the Roman chicken and her egg found their way into German pamphlets and broadsides.

The Comet Egg of 1680 was said to have been the first effort by a Roman hen during the time the Great Comet of 1680 was in the sky (from the collection of D. K. Yeomans).

Silver medal struck in 1681 to commemorate the appearance of the Great Comet of 1680–81 (from the collection of D. K. Yeomans).

THE GREAT COMET OF 1680

During the seventeenth century, the public's susceptibility to superstitious fear of comets increased. This trend began to reverse toward the end of the century by the publication of Isaac Newton's work on the Comet of 1680 and Edmond Halley's work on the Comet of 1682. The Comet of 1682 would eventually bear the name of Halley.

However, before the work of Newton and Halley could reverse the rising tide of cometary superstition, the Comet of 1680 would witness a tidal wave of foolishness. Of the approximately 208 known broadsides on comets, 62 refer to the Comet of 1680. (The front page of the *National Enquirer* is the modern equivalent of a seventeenth century broadside.) In Germany alone 100 treatises were published on this comet. Of these treatises, only four were written to quiet superstitious fears. The authors of these four works must have been offering unpopular views because three of these works were published anonymously, and the other was signed by the author's initials only.

In addition, at least a dozen medals were struck to commemorate the comet's impressive appearance in late 1680 and early 1681. The German medal shown here has on one side an illustration of the comet on a stellar background with the inscription Ao 1680 16 December–1681 January. On the other side is an inscription in German that roughly translates to "The star threatens evil things; trust in God who will turn them to good." The capitalized letters in the German legend make a chronogram when rearranged into roman numerals (i.e., MDCLVVVVVVI in arabic numerals is 1681.)

farming, however. He spent most of his childhood reading and building strange mechanical toys. His family finally sent him to Trinity College at Cambridge University in 1661. Newton did not distinguish himself there either.

In 1665, a terrible epidemic of bubonic plague broke out in London and began to spread across the country. Cambridge closed its doors and sent all its students home. The plague lasted a year and a half. During that period, working all by himself at home in the village of Woolsthorpe in east–central England, Newton

A French caricature by Honoré Daumier poking fun at the foolishness that followed an incorrect prediction of a comet striking the Earth in 1857 (photograph courtesy of P. Veron).

AN ASTRONOMICAL CHARLATAN

The discoverer of a comet receives instant fame and recognition. He is assured a place in history by having his name forever associated with the comet. The rewards of a comet discovery were apparently too much of a temptation for the Chevalier Jean Auguste D'Angos (1744–1833). D'Angos was a captain in the French army and formulated the law of gravity, began to devise a new mathematical tool called calculus,[3] made a major breakthrough in optics, and investigated the nature of light.

When Cambridge University reopened in 1667, Newton returned to school with his new ideas. His professor, Isaac Barrow, was so impressed that he resigned his professorship in 1669 and had Newton appointed in his place. Newton was twenty-seven years old.

By that time, Newton had built the first reflecting telescope.

became a member of the Knights of Malta. He was also a physician, chemist, and astronomer. In 1783, the Grand Master of the Knights of Malta invited D'Angos to Malta, and a fine observatory was built for him there.

One year after occupying his fine observatory, D'Angos sent the famous French astronomer Charles Messier a letter announcing the discovery of a comet in the constellation Vulpecula. D'Angos also sent Messier two position observations of the alleged comet as well as an orbit that he had computed from his observations. Despite repeated attempts, neither Messier nor others could find the new comet. Nor could anyone reproduce D'Angos' computed orbit. When pressed for more details, D'Angos reported that all the observations had been lost when the observatory burned down. (D'Angos' experiments with chemistry, in particular phosphorus, resulted in a disastrous fire that destroyed the observatory and its records only five years after it was built.) Subsequently, a German periodical of 1786 that listed fourteen observations of the comet as given by D'Angos was discovered.

The German astronomer Johann F. Encke began an investigation, and in 1820 he found that the orbit computed by D'Angos was consistent with D'Angos' fourteen observations only if Encke, in each case, used an Earth–comet distance that was exactly ten times the correct value. D'Angos had fabricated his observations of this nonexistent comet by (incorrectly) using a bogus orbit he had invented. D'Angos also "discovered" bogus comets in 1793 and 1798. Encke noted that "D'Angos had the audacity to forge observations that he never made of a comet that he had never seen, based on an orbit that he had gratuitously invented, all to give himself the glory of having discovered a comet."

EDMOND HALLEY: THE MAN

Who was the first to design and test deep-sea-diving bells and helmets? Who proposed and commanded the first naval expedition for scientific research carried out by the British government? Who discovered that half the people of the Polish city of Breslau were dying by the age of seventeen and then calculated the first mathematical tables of life expectancies of populations? Who created the modern method of mapping trade winds and plotting lines of magnetic variation? The answer in every case is Edmond Halley.

Today, most people know of Edmond Halley only because a famous comet bears his name. But during Halley's career, his work on comets was viewed as a minor achievement in a lifetime of great successes.

Except for Isaac Newton, Edmond Halley was the most distinguished scientific figure of his time. During the eighty-five years of his life, Halley (1656–1742) contributed significantly to meteorology, mathematics, and navigation and laid the foundations

Edmond Halley at the age of eighty. Frontispiece from Halley's Astronomical Tables, *London, 1752 (from the collection of D. K. Yeomans).*

of physical geography, geophysics, geomagnetism, physical oceanography, and life insurance statistics.

But Halley's greatest efforts were reserved for astronomy, and his achievements range far beyond the comet named in his honor.

In astronomy, Halley was the first to recognize that each star has a motion of its own with respect to the other stars (its proper motion); the first to suggest that many nebulas (hazy objects seen between the stars) are vast clouds of interstellar gas glowing by processes within them; the first to propose that some stars vary in brightness because they themselves are changing their light output; and the first to apply the concept of gravity to the universe at large. Throughout his life, Halley contributed greatly to the most difficult practical astronomical problems of his day: finding the distance to the Sun (from which all the planetary distances could be computed) and calculating longitude while on board a ship.

Edmond Halley was born in London on October 29, 1656. His father was a prosperous soapboiler (soapmaker), salter (one who salted meat to preserve it, particularly for sea voyages), and landowner. He encouraged his son's interest in astronomy with schooling and instruments.

Halley published his first scientific paper at the age of nineteen and at age twenty he dropped out of Oxford to take the telescope and sextant his father gave him south of the equator to compile the first serious catalog and chart of the stars of the southern skies

*Edmond Halley
(photograph courtesy of Pulkovo
Observatory).*

which can never be seen from Europe. His work was so useful and skillfully done that, upon his return, Oxford gave him a master's degree without requiring him to take any exams.

He went on to be, among many things, the first Clerk of the Royal Society (the great English organization of scientists) and gathered and published the works of its members. He was for three years and three separate voyages appointed a Captain in the Royal Navy (without any shipboard training) and given full command of a ship to research longitude measurements, magnetic variation, and land mapping. He was appointed Savilian Professor of Geometry at Oxford in 1704, and was named England's second Astronomer Royal in 1720.

Halley possessed rare scientific ingenuity, as was evident when he was asked in 1693 to calculate the total acreage in each of the counties in England and also the country's total land area, a problem that might have involved years of surveying. Instead, Halley took the biggest and most accurate map of England available and drew the largest circle possible within its land mass. The radius of this circle was 69$^{1}/_{3}$ miles, and from this measurement Halley calculated that this circular region had within it 9,665,000 acres. He cut out this section of the map and weighed it. He then cut out the entire area of England and weighed it. The map of all England weighed four times as much as the map of the circle, so he calculated that the land area of England was 38,660,000 acres, an estimate that we know today is too high by only slightly over three percent. To calculate the county acreages, Halley simply cut out each county from the map and weighed it in proportion to the circle whose acreage he knew.

What kind of human being was this exceptionally energetic, wide-ranging, and incisive scientist? Although few of his personal papers have survived, he seems to have led a full and happy private life. He was married to Mary Tooke for fifty-five years, and they had three children: two girls and a boy.

Whenever a job required enormous diplomatic, organizational, or administrative skill, his friends in the Royal Society (many of the best minds in England) turned to him. Halley was a close friend to both Robert Hooke and Isaac Newton, two very difficult personalities who thoroughly disliked one another. Nearly everyone found Halley to be a man of warmth, honesty, tact, openness, and humor.

Even sour, paranoid John Flamsteed, the first Astronomer Royal, who turned against Halley viciously by urging officials not to grant Halley an Oxford professorship because he would "corrupt ye youth of ye University with his lewd discourse," also confided that ". . . were he ether [either] honest or but civil there is none in whose company I could rather desire to be."

The extraordinary humanity of Edmond Halley could be seen in his relationship with Isaac Newton. It was Halley in 1684 who first recognized what Newton had achieved in his formulation of gravity, and it was he who persuaded Newton to publish. In fact, Halley used his own money to publish Newton's *Principia* (1687). Because of Halley's efforts, Newton was recognized as the foremost scientist in the world. Yet, Halley never showed a moment of jealousy or resentment. Augustus De Morgan, an English mathematician of the nineteenth century, said of Halley's work and generosity on behalf of Newton and the *Principia*: "But for him, in all human probability, the work would not have been thought of, nor when thought of written, nor when written printed."

A sidelight to this story also sheds light on Halley's character. Not only was Halley never reimbursed by the Royal Society (or Isaac Newton) for publishing Newton's revolutionary *Principia*, but the Royal Society was not even paying Halley his meager annual salary of fifty pounds for his services as Clerk (publications editor and corresponding secretary). The Royal Society had recently exhausted its treasury by publishing *Historia Piscium (History of Fishes)* by Francis Willughby. The Society would have been rich if this book had sold well, but this expensive volume moved at the rate of a sea anemone. So, although Halley had no role in the bad decision to publish this book, the Royal Society did not pay Halley in 1687 or 1688. Instead of a salary, they gave Halley fifty copies of the virtually unsalable fish book, and then they gave him another twenty-five copies because of how far his salary was in arrears.[1]

Yet, through it all Halley went right on doing a superior job for the Royal Society.

Curiosity, energy, integrity, and personal warmth were Halley's hallmarks throughout his long and illustrious life.

Edmond Halley's health began to decline at the age of eighty because of a series of strokes, but his mind remained acute to the end. He died on January 14, 1742: ". . . being tired . . . he asked for a glass of wine, and having drunk it presently expired as he sat in his chair without a groan. . . ."

[1] Halley probably tried to place *Fishes* with booksellers. In late 1690, the Royal Society finally paid Halley the two years' salary due him for 1687 and 1688. They did not add any interest, unless you count *Fishes*.

J. G. Palitzsch. From Das Weltall, *Vol. 7, April 1, 1907, facing p. 189.*

When the Royal Society heard about it, they asked to see the instrument. Newton supplied a new and larger model and was elected a Fellow of the Society. He then sent the Royal Society a paper showing how, by the use of prisms, he had proved that white light is a mixture of all the colors of the rainbow. But his penetrating insight into the nature of color was greeted with skepticism, especially from Robert Hooke. Newton, sensitive to criticism, withdrew into his research at Cambridge. For the next twelve years Newton worked quietly, and no one knew the scope and brilliance of his discoveries.

In August 1684, Halley visited Newton at Cambridge University. As later reconstructed by an acquaintance of theirs,[4] Halley explained the problem of planetary motion that had frustrated Hooke, Wren, and himself. A force kept the planets traveling around the Sun, a force that grew weaker with distance. This force caused the planets to travel faster when they were closer to the Sun and slower when they were farther away. In fact, at twice the distance, the force was only one-fourth as great. But how could such a force be proved, that is, demonstrated and described precisely by mathematics? "None of us can do it," said Halley. "Can you help?"

"I have calculated it," said Newton. "I worked it out seventeen years ago."

Halley was stunned. "How?" he asked. "Where? Can I see?"

"Well," said Newton, "I wrote the full solution down five years ago. It's around here somewhere." And he shuffled through some papers.

"You can't be serious," said Halley. "You explained the force that controls the planets, and you never told anybody about it?"

"It explains the moon in orbit around the Earth and the tides also," Newton added. "And how dropped objects fall to the ground. Let's see. I think I put it in this drawer. No, not here either."

"It's lost?" groaned Halley.

"Yes, but no matter," said Newton. "I'll work the proof out again and send it to you in London as soon as possible."

Three months later, a messenger brought Newton's calculations to Halley. They were the nucleus of the law of universal gravitation.[5]

Halley quickly recognized that Newton was as great a scientist as had ever lived. He went to visit Newton again and persuaded him to write down more of his ideas and discoveries. Halley anticipated a small book that would be published by the Royal Society.

Newton worked for eighteen months on the project. The little book grew into a huge volume. The Royal Society did not have

*A nineteenth century French cartoon showing
a comet colliding with the Earth.*

WILLIAM WHISTON AND COMET–EARTH COLLISIONS

Even after the work of Newton and Halley showed that comets
were celestial objects obeying the same universal law of gravity that
governed the motions of other celestial bodies, the fear of comets
persisted. Halley had shown that the comet that now bears his
name moved in a highly elongated ellipse—a path that brought the

comet back to the inner solar system approximately every seventy-six years. It seemed reasonable to suppose that other comets had similarly elongated paths and that many could cross the Earth's orbit on their way toward the Sun and again as they traveled away from the Sun. If the Earth and comet met at an intersection, it would be no mere evil portent for a king. The collision would be a full-scale disaster for everyone. In the eighteenth century, comets were no longer thought of as signs of God's wrath; they were the agents of God's wrath.

The popular fear of a cometary collision was at least partially attributable to William Whiston, an English clergyman and mathematician. Whiston, who was named by Isaac Newton as his successor at Cambridge University, published his ideas in 1696 in a book entitled *A New Theory of the Earth.* By an impressive manipulation of Newtonian science and biblical chronology, he explained Noah's flood and the future end of the world by close approaches of a comet to the Earth. According to Whiston, the deluge was initiated by the near miss (within 16,000 miles) of a comet on either November 28, 2349 B.C. or December 2, 2926 B.C. The exact date depended on whether one chose to believe the more modern Hebrew sources or the ancient Samaritan texts. The flood was caused in part by the comet raising an enormous tide on the Earth's surface and subsurface waters. According to Whiston, the comet's "watery atmosphere" also touched the Earth and "thus were opened all the cataracts of heaven." The end of the world would come when another comet came close to the Earth, altered our planet's orbit, and caused it to pass too close to the solar inferno. Alternatively, Whiston mentioned that the end of the world might come even sooner if the comet actually collided with the Earth.

Originally Whiston had no particular comet in mind. However, when Edmond Halley (erroneously) suggested that the comets seen in A.D. 1680, A.D. 1106, and 44 B.C. were the same comet with an approximate period of 575 years, Whiston amended later editions of his book to specify this comet as the cause of the biblical flood and the future end of the world. The world would end, he announced, at this comet's next return in A.D. 2255. Whiston's book was well received (the sixth edition was published in 1755) and helped establish the notion that a cometary collision with the Earth was a distinct possibility.

enough money to print it, so Halley paid all the printing costs himself. *Philosophiae Naturalis Principia Mathematica (The Mathematical Principles of Natural Philosophy)* by Isaac Newton was published in 1687. It forever changed the course of science.

"IF IT SHOULD RETURN"

Halley was fascinated that Newton's concept of gravity could simultaneously explain how apples fall from trees on Earth and how planets revolve around the Sun in space. He remembered also the comet he saw in 1682.

"THE FERRET OF COMETS"

"The Ferret of Comets" was the tribute that French king Louis XV paid to Charles Messier (1730–1817). Messier's fervor for comets was kindled in part when, as assistant to the Astronomer of the Navy, he became the first man in France to find Halley's Comet on its first predicted return in 1758–59.

Messier, without formal astronomical training, devoted the rest of his life to finding comets. So that nothing in the sky would deceive him, he gradually collected a list of 103 objects that were hazy in appearance through a small telescope and thus might be confused for comets. Messier was not much interested in these distractions, but his catalog, the first of its kind, became a gold mine for astronomers who wanted to study these clouds of interstellar gas and dust (nebulas), star clusters, and galaxies.

Larger catalogs of nebulous objects were to follow (even within Messier's lifetime), but even today both professional and amateur astronomers often refer to the brighter nebulous objects, star clusters, and galaxies by their Messier numbers: the Crab Nebula (a supernova remnant) as M 1, and the Great Spiral Galaxy in Andromeda as M 31.

Messier's dedication to the discovery of comets was legendary and gave rise to frequently repeated tales at his expense. One story

Was it possible that gravity ruled the motion of the comets, too?[6] The idea that comets travel in orbits around the Sun had been proposed before, but no proof had ever been offered.[7] Great scientists such as Kepler had argued that comets travel in straight lines—vagabonds passing through our solar system.

Halley gathered all the data he could find on comets. Only twenty-four comets had ever been observed with enough care to be useful to his project. He set to work to discover whether comets were affected by the gravity of the Sun.

The calculations done with Newton's new mathematical

told that when Messier missed some observing time while nursing his dying wife, his rival Montaigne discovered a comet. At his wife's funeral, a friend offered condolences on his recent loss. "Alas," said Messier, "Montaigne has robbed me of my thirteenth comet."

The story is almost certainly untrue. Montaigne discovered comets in 1772, 1774, and 1780, but Messier's wife, to whom he was quite devoted, died in 1798. By 1798, according to contemporary count, Messier had discovered twenty-one comets.[1]

Another story relates that Messier used every available moment to watch for comets, even when no telescope was at hand. One evening in 1781, while walking through a novelty landscape garden full of pagodas, pyramids, castles, tombs, and windmills, Messier was staring at the sky so intently that he fell down a well.

Actually, Messier found what he thought was a little grotto with its door open, so he stepped inside. The door, left open by mistake, stood at the top of an underground ice storage cellar, and Messier fell twenty-five feet onto the ice below. He broke many bones and barely survived. One wrist and leg remained deformed. He could not resume his comet searches for a full year.[1]

What is true in the distorted stories about Messier is that he singlemindedly (some would say obsessively) pursued comets. Today Messier is credited with discovering thirteen or fourteen comets. Only Jean Louis Pons discovered more.

[1] See Kenneth Glyn Jones: *Messier's Nebulae and Star Clusters* (New York: American Elsevier, 1969), pp. 394–98.

"THE COMET MAGNET"

The record for most comets discovered is held by Jean Louis Pons (1761–1831).

Pons had no interest in astronomy before becoming the doorkeeper of the Marseilles Observatory at the age of twenty-seven. From the director and astronomers there, he gradually absorbed information and discovered his first comet in 1801 at the age of forty. From then until his death in 1831, he discovered a total of thirty-seven comets (some say thirty-six because of other people's claims to have seen certain comets earlier). Twice, in 1808 and 1826–27, he discovered five comets in less than one year. In a twenty-five-year period, he was the discoverer or codiscoverer of three-fourths of all the comets found. No wonder he was called the "comet magnet."

Pons could find comets, but, unlike Messier, he did not observe their positions accurately. His position measurements for three of the comets he found in 1808 were so poor that no orbits for these comets could be calculated.

Pons was always asking noted astronomers for ideas to help him find more comets. One time, Hungarian astronomer Franz Xaver von Zach decided to tease him by telling him to look for comets when there were many sunspots. Pons took the advice, and when sunspot activity increased, Pons promptly found another comet and cabled his gratitude to von Zach. Von Zach respected Pons enough to help him obtain the directorship of two observatories.

methods were some of the most difficult and time-consuming ever performed in the history of astronomy. Begun in 1687, Halley's results were not ready for eighteen years, until 1705. They showed unmistakably that comets respond to the gravity of the Sun.

But something else about his twenty-four comets struck Halley even more forcibly. Three of these comets, the ones reported in 1531 and 1607, and the one he observed in 1682, all traveled almost identical courses around the Sun. And they were seen an average of seventy-six years apart. They could be three different comets following one after another, or these three apparitions could be just one comet returning to the Sun approximately every seventy-six years.

That was the answer. And that meant that the Comet of 1682 should return in seventy-six years. Halley wrote:

> I may, therefore, with some confidence, predict its return in the year 1758. If this prediction is fulfilled, there is no reason to doubt that other comets will return.[8]

On Christmas night in 1758, the great comet was sighted by Johann Georg Palitzsch, a farmer and skillful amateur astronomer in Germany. Edmond Halley did not live to see it. He had died sixteen years earlier.[9] But ever since, this most famous of comets has been known as Halley's Comet, named for the man who proved that comets obey the laws of physics and travel in orbits around the Sun.

These appearances in 1531, 1607, and 1682 were not the only times that Halley's Comet had been reported. The Comet of 240 B.C. that appeared near the height of Greek astronomical discovery was Halley's Comet. And so were the Comet of A.D. 451 when the Huns were repulsed and the brilliant Comet of A.D. 837. Halley's Comet appeared just before the Battle of Hastings in 1066 and was recorded on the Bayeux Tapestry. Probably the appearance of Comet Halley in 1301 is what we see in Giotto's painting, "The Adoration of the Magi." And Halley's Comet in 1456 paralyzed the Turkish and Catholic armies before the Battle of Belgrade.[10]

Today we know that Halley's Comet has been observed and recorded on all of its thirty visits to the Sun since 240 B.C.[11]

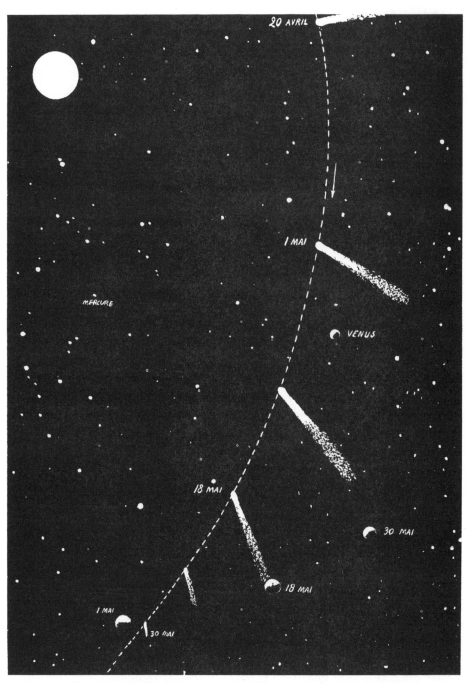

On May 19, 1910, Halley's Comet passed between the Sun and Earth
and close enough to our planet so that part of its tail, extending away
from the Sun, grazed the Earth. No effects were observed. From
L'Illustration, *Vol. 135, Jan. 22, 1910, p. 60 (photograph courtesy of
Ruth Freitag).*

Chapter Three

1910: THE TALE
OF A TAIL

> *I came in with Halley's Comet in 1835.*
> *It is coming again next year, and I*
> *expect to go out with it.*
> —MARK TWAIN

LAST TIME AROUND

By 1910, more than 200 years had passed since Edmond Halley had demonstrated that comets travel in orbits around the Sun and obey the law of universal gravitation that Newton had discovered.[1] Fear of comets and superstitions about comets should have been things of the past, but they weren't.

What made the Earth's encounter with Halley's Comet in 1910 so interesting and unusual was that our planet was positioned so that the comet passed precisely between the Earth and the Sun. The Earth at that time was not only at its closest to Comet Halley during the 1910 apparition, but, as the two swept by one another, traveling in opposite directions, the comet passed directly in front of the Sun.[2] The calculated date for this event was May 18, 1910, as seen from islands in the Pacific. Astronomers tried hard to detect Comet Halley as a dark dot moving across the Sun's bright face so that they could measure the comet's size. If it were as big as 50 miles (80 kilometers) in diameter, they were confident they would see it. They didn't. The estimated size of comets had to be revised downward.

*Comet Morehouse (1908 III). Cyanogen, a poisonous gas, was
discovered in its coma by spectroscopy. The prospect of
cyanogen poisoning the Earth in 1910 when the tail of Comet
Halley brushed our planet caused concern among
nonscientists (photograph by E. E. Barnard courtesy of
Yerkes Observatory).*

Yet there was another fascinating aspect to the comet's passage
between the Earth and Sun. Because a comet's tail always flows
outward, away from the Sun, the tail of Comet Halley would
extend toward the Earth.

But would the tail extend far enough to envelop our planet?
Comet Halley routinely develops a tail 30 million miles (50 million
kilometers) long when near the Sun. On the night of May 18–19,
1910, the nucleus of Halley's Comet would be 14 million miles (23
million kilometers) from Earth. The Earth would pass through at
least a portion of the comet's tail.

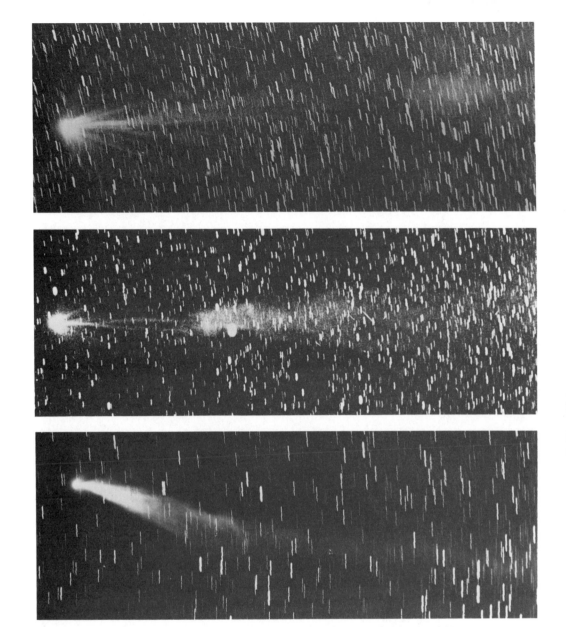

These views of Comet Morehouse on Sept. 30, Oct. 1, and Oct. 2, 1908, show how variations in the solar wind can give the appearance of carrying off a comet's tail while the comet grows a new one (photograph courtesy of Yerkes Observatory).

In the middle of the nineteenth century, astronomers began to use spectroscopes to study the light from comet comas (heads) and tails to determine their chemical composition. In 1908, the coma of Comet Morehouse showed the presence of cyanogen (C_2N_2), a poisonous gas.

OUT OF THE WOODWORK: PUBLIC REACTION TO HALLEY'S COMET IN 1910

Despite assurances from scientists in 1910 that the nucleus of Halley's Comet posed no threat to Earth (it would never be closer than approximately sixty times the distance of the Moon) and despite assurances that the cyanogen in the comet's tail was minute in quantity and would be quickly converted to harmless chemicals in the Earth's thick atmosphere, newspapers throughout the world recounted numerous stories of public panic.

People in Lexington, Kentucky and many other places flocked to all-night religious services "to prepare themselves to receive the celestial visitor and meet their doom."

In eastern Texas, logging operations closed down because the lumberjacks fled to revival meetings.

In Nebraska and Wisconsin, people tore the lightning rods off the roofs of their houses and barns for fear that metal would attract the comet.

How best to face the threat of the comet varied. Thousands of coal miners in Pennsylvania refused to enter the pits because they wanted to be on the surface when the world ended. Miners in Colorado, however, wanted to stay underground in hope of surviving the comet peril.

Others coped less well. A young secretary in New York City stayed up all night to see the comet. She was so disturbed by the sight that she lost her memory and was found by police wandering the streets in a daze.

A sheepherder near Walla Walla, Washington, read so much about Halley's Comet that he became disoriented and began to rave. He was confined to a padded cell. Another sheepman near San Bernardino, California, erected a cross and nailed both his feet and one hand to it before he was rescued from his do-it-yourself crucifixion.

Burglar (with sudden enthusiasm for astronomy): "Scuse me, Guv'ner, can you tell me where I can get a view at this 'ere comet?"
Cartoon by Lewis Baumer. From Punch, *Vol. 138, June 1, 1910, p. 403 (reproduced with permission from Punch Publications).*

Because the gases in the coma flowed out into the tail and because the tail of Halley's Comet would brush our planet, was life on Earth in jeopardy?

Newspaper articles from 1910 show that many scientists hastened to assure the public that Halley's Comet would not crash

OUT OF THE WOODWORK—*cont.*

Sadly, people in New York, Pittsburgh, Denver, and San Francisco, "crushed by a premonition of danger," committed suicide.

In Montreal, a gust of wind caused a door to slam shut. The woman of the house, an invalid, leapt to her feet crying "The comet has struck!" then crumpled to the floor and died.

In Newark, New Jersey, police arrested Luigi Ciefice under suspicion that he was a member of a crime ring called the Black Hand. In jail, the man read in the newspapers about the Earth's impending brush from Comet Halley's tail. Ciefice fell to his knees, praying, then yelled, "All the people will be killed. I will be killed. I cannot die without confessing that I killed Patrick Cahill. I put him in the ground behind my house." The police did not know that a murder had been committed, but the story was verified.

Yet "protection" against the comet fumes was available, for a price. Expensive "comet pills" sold briskly in New York (one dollar each) and south Texas. In Haiti an enterprising and increasingly wealthy voodoo doctor was dispensing comet pills as fast as he could make them to ships' crews, tourists, and natives. In Atlanta, street peddlers hawked "conjur bags" as protection against the comet. In South Africa, one man bricked up his living room, put in a supply of oxygen cylinders, then advertised for a limited number of people to rent his comet shelter.

Halley's Comet proved to be quite a boon to crooks. Two well-dressed young men drove into the farming community of Towaco, New Jersey, on April 28, 1910. They announced that they were scientists; that Halley's Comet would be closest to Earth at 3:00

into the Earth, that comet tails are extremely tenuous (more rarefied than any vacuum yet made on Earth), and that the Earth's thick atmosphere offered excellent protection.

Not everybody was convinced. It didn't matter that the quantity of cyanogen in the comet's tail was probably minute. It

A.M. the next morning; and that they were offering ten, five, and two-and-one-half dollar gold pieces to the three people who could provide the best description of the comet. By 2:00 A.M. on April 29, everybody in Towaco was on top of nearby Waukhaw Mountain. On April 29, the comet was not closest to Earth; in fact, it was too nearly behind the Sun to be visible. When the disappointed Towacoans got home, they found that almost all their chicken coops had been raided.

Here and there, too, practical jokers were at work. The citizens of Rochelle Park, New Jersey, were already a bit on edge because the Earth would plunge through the comet's tail that evening. A man and his son, for a little amusement, released a balloon carrying dynamite, sodium, and a fuse. The balloon rose into the sky, exploded, and sent the burning sodium outward in a shower of flame. The community was terrified.

Most people laughed at the doomsayers and treated the comet as a welcome, rare, and interesting sight. Throughout the United States, Britain, and France, people held "comet parties," sometimes in costume and sometimes on the roofs of buildings so that the comet could be seen in the early morning hours.

One New York bar invented the "cyanogen cocktail"; another invented the "syzygy fizz." (A syzygy is a lineup of three celestial objects; in this case, it referred to the passage of Halley's Comet directly between the Earth and the Sun.) Still another bar described its "comet cocktail" as "a seething concoction made with cracked ice, a snifter of French vermouth, and a jigger of apple jack. Six of these, after being well shaken, are guaranteed...to make even a blind man see the comet."

SPECTRAL LINES AS CHEMICAL FINGERPRINTS

Because of its wavelike nature, the propagation of light is analogous to traveling ocean waves. Although ocean waves can move along quite rapidly, the underlying water itself moves only up and down like a bobbing cork. However, unlike ocean waves or sound waves, light travels at the fastest possible speed, and it can travel through a complete vacuum. The distance between successive wave crests is called the wavelength. The wavelength of light determines visible colors. Visible light wavelengths are so small that a special unit is used to describe them. The angstrom, named after the nineteenth century Swedish physicist Anders Angstrom, is one 100-millionth (10^{-8}) of a centimeter in length. The human eye is sensitive to the wavelength region ranging from approximately 4,000 to 8,000 angstroms.

An imaginary surfer riding the crest of a light wave of 5,000 angstroms would find himself sandwiched between wavecrests only five ten-thousandths of one millimeter behind and in front of him, more than 1,000 times smaller than the period at the end of this sentence. Not only would our surfer be crowded, but he would be moving at the speed of light—300,000 kilometers (186,000 miles) per second. At this rate, he could circle the Earth in slightly more than one-tenth of one second.

If our eyes could see beyond the visible wavelength region to shorter wavelengths, we would see ultraviolet radiation between 100 and 4,000 angstroms and X-rays between 1 and 100 angstroms. Our imaginary surfer might feel more at home with infrared wavelengths of approximately one millimeter or perhaps radio wavelengths between one centimeter and several meters. But whatever wavelength he chose, our intrepid surfer would still be traveling at the speed of light.

Isaac Newton first suggested that a glass prism separated white sunlight into its component colors, the rainbowlike spectrum. In 1859, the German physicist Gustav Kirchhoff finally provided the

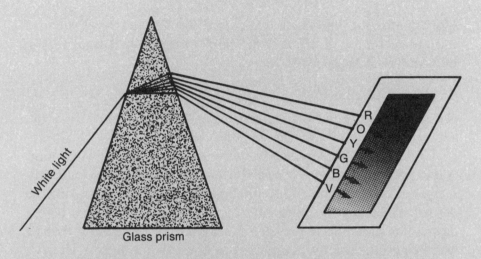

A prism breaks white light, made up of many colors, into its component colors of spectral lines.

key to identifying the chemical makeup of an object from its observed light. Simply stated, Kirchhoff noted that a luminous solid or liquid will emit light of all wavelengths—a continuous spectrum. A rarefied luminous gas (such as the coma of a comet) will emit visible light, and its spectrum will show individual bright lines. These same bright lines will appear as dark lines if the source of a continuous spectrum is viewed through a rarefied luminous gas that is cooler.

Each element or molecule in a star or comet will reveal its identity by its own peculiar set of spectral lines; each has an identifying fingerprint. From laboratory experiments and theoretical calculations, a file of spectral fingerprints has been developed for many elements and molecules. When the spectrum of a comet is observed either in the ultraviolet, visible, infrared, or radio region, the spectral lines from the gases in the comet's coma and tail can often be identified by using the spectral fingerprints on file, and a new cometary constituent can be identified. Often a comet will show a continuous spectrum due to sunlight reflected from the abundant dust particles surrounding the comet.

didn't matter that the Earth had passed through the tail of a comet in 1861 with no ill effects.[3] It didn't matter. To some people, disaster was at hand.[4]

CURIOUS BEHAVIOR

Newspapers recorded people in Chicago plugging up their doors and windows to keep out the noxious vapors.[5]

A man in Somerville, New Jersey, who had always been "sober and industrious," brooded over the immersion of the Earth in Comet Halley's tail and concluded that the world would be destroyed. Several mornings later, a policeman spotted him before dawn running along the street "in scant attire," screaming that he was being pursued by his mother-in-law and the tail of Halley's Comet.[6]

Thirty miles south of Aline, Oklahoma, said one report, Henry Heineman led his cult, the Select Followers, in a grisly ritual to ward off the comet. He was about to plunge a knife into sixteen-

1910 French postcard (from the collection of D. K. Yeomans)

1910 cartoon (photograph courtesy of J. Dudley, Royal Greenwich Observatory)

year-old Jane Warfield as a blood atonement for the sins of mankind. Heineman said that God had commanded him to perform this sacrifice because otherwise "the world would end and the heavens would be rolled up like a scroll following contact with the tail of the Comet." A sheriff's posse arrived just in time to prevent the girl's death.[7]

In southern Texas, two salesmen and their agents went from town to town selling phony "comet pills" and leather gas masks to protect people from the dangerous vapors. Analysis showed the pills were made of sugar with just enough quinine to make them bitter just as people expected medicine to be. The salesmen were arrested. But before they could be tried, a crowd of angry citizens gathered outside the courthouse. They refused to testify against the salesmen and demanded that the swindlers be released so that they could buy comet pills and "inhalers" before it was too late. Apparently the sheriff gave in to their demands.[8]

PUTTING HALLEY'S COMET TO WORK

by **Ruth S. Freitag**
Senior Science Specialist
Library of Congress

Halley's Comet in the spring of 1910 was a media event. Comet stories vied for newspaper space with reports on the travels of ex-President Teddy Roosevelt, the floods in Paris, the eruption of Mount Etna, the devastating earthquake in Costa Rica, and the deaths of Mark Twain and King Edward VII. Indeed, the comet was often blamed for these sensational occurrences.

The comet inspired reams of newspaper verse, comic cartoons and postcards, jokes, short musical compositions, and revue sketches. The Ziegfeld Follies of 1910 featured a song called "The Comet and the Earth." A cartoon in the *Portland Oregonian* suggested that the nation's professional funnymen should offer the comet a vote of thanks. Today Halley's Comet T-shirts are being sold to celebrate the comet's return, but in 1910 comet designs decorated vests, neckties, handkerchiefs, and socks.

The manifestation of interest in the 1910 visit of Halley's Comet that is perhaps least familiar today was the application of the comet motif to advertisements. A sweeping arc of tail, growing out of a bright circular or star-shaped head that usually featured the product, formed an eye-catching design that was used to good effect by many commercial artists.

Because libraries often remove advertising sections from periodicals before binding them to lessen the bulk of the volumes, many such advertisements have perished; others are preserved only in the pages of newspapers that have ben reduced to microfilm, from which it is difficult to reclaim clear, full-size images. That they were quite numerous is implied by the cartoon "Publicité céleste" ("Celestial Advertising"), showing a bemused comet surrounded by a host of imaginative appeals.

(Condensed by permission from
The Quarterly Journal of the Library of Congress,
Summer 1983)

Ads using comet motifs (left to right: from Canadian Magazine, *Vol. 34, April 1910; from* Simplicissimus, *Vol. 15, Apr. 18, 1910, p. 50; from* Saturday Evening Post, *Vol. 183, Feb. 19, 1910, p. 47; from* Lookout, *Vol. 5, Apr. 30, 1910; from* Illustrirte Zeitung, *Vol. 134, Mar. 3, 1910, p. 371; from* Wide World Magazine, *Vol. 25, 1910)*

Ads using comet motifs (left from Sketch, *Vol. 70, Apr. 13, 1910; right from* Sphere, *Vol. 41, May 14, 1910).*

THE WAY TO GO

The prospect of Comet Halley's tail enveloping the Earth gave rise to many quaint predictions.

Because a comet must contain much hydrogen, reasoned one astronomer, as the comet's tail swept across the Earth, the hydrogen in the tail could mix with the oxygen in the air and set off a tremendous worldwide atmospheric explosion as the hydrogen and oxygen combined to form water,[9] "followed instantly by a deluge of [rain] and leaving the burnt and drenched Earth no other atmosphere than the nitrogen now present in the air, together with a relatively small quantity of deleterious vapors."[10]

A happier end of the world was proposed by another speculator who noted that a comet's coma contains cyanogen, made of carbon and nitrogen. The comet's nitrogen, he said, would combine with the Earth's oxygen to form nitrous oxide (N_2O), known as "laughing gas." The people of the Earth would not survive, but they would go out laughing. Earthlings would all "dance, deliriously happy, to an anesthetic death."[11]

Among the many people fascinated by the return of Halley's Comet in 1910 was the great American writer Mark Twain. He had an avid interest in astronomy. One morning in 1909, after he had calculated how far light would travel in one year (almost 6 trillion miles—a light year), he turned to his friend Albert Bigelow Paine and said:

> I came in with Halley's Comet in 1835. It is coming again next year, and I expect to go out with it. It will be the greatest disappointment of my life if I don't go out with Halley's Comet. The Almighty has said, no doubt, "Now here are these two unaccountable freaks; they came in together, they must go out together." Oh! I am looking forward to that.[12]

Halley's Comet passed closest to the Sun on April 20, 1910.[13] Mark Twain died the next day.

On the night of May 18–19, 1910, as Comet Halley sped outward from the Sun, the Earth plunged through its tail.[14] Nothing happened. Bright moonlight prevented astronomers from seeing if there might be a glow in the sky. No extra meteors were counted.

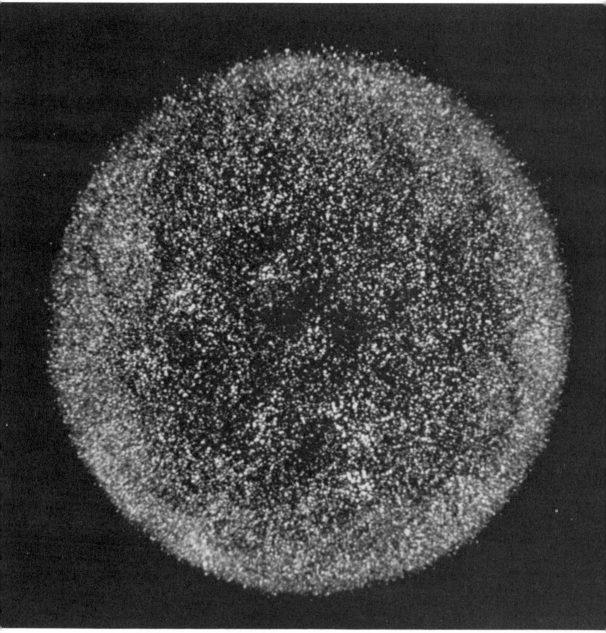

*The spherical Oort Cloud of comets surrounds the planetary
realm of the solar system (photograph courtesy of the Science
Museum of Virginia).*

Chapter Four

ONCE AROUND
WITH HALLEY:
A COMET'S PERSPECTIVE

No more politicians,
No more tariff schemes,
No more trust conditions,
No more quick-rich dreams!
Bang! Annihilation!
Smash! We fly to bits!
There's some consolation
If the comet hits!
—PAUL WEST, "If the Comet Hits"[1]

COASTING OUTWARD

In 1910, Halley's Comet had once more kept its rendezvous with the Sun. For the previous thirty-eight years, one-half of its orbital period, it had fallen toward the Sun, gathering speed under the influence of gravity. In 1910, Halley's Comet rounded the Sun at almost 122,000 miles (196,380 kilometers) per hour, and its momentum carried it outward for the next thirty-eight years.

Imagine that we could have followed Halley's Comet outbound in 1910 and traveled with it for a lifetime until it returned.

Under the steady tug of the Sun's gravity, the comet was losing speed. Its present orbit would never carry it to the stars or

THE COMETARY NUCLEUS

Until spacecraft can take close-up pictures of a comet's nucleus, all the information that we have on the nucleus is inferred and not directly observed. Even in the largest telescopes, the cometary nucleus is too small to observe as anything other than a point of light. Comet Halley is estimated to be approximately 6 kilometers (3.7 miles) in diameter. Although small, cometary nuclei must be capable of generating the prodigious amounts of gas and dust that are observed to surround comets, and they must be strong enough to survive for hundreds of close encounters with the Sun.

In 1950, Fred L. Whipple of the Smithsonian Astrophysical Observatory in Cambridge, Massachusetts, outlined a model for comet nuclei that has withstood the test of time. Whipple suggested that the nucleus, the heart of all cometary phenomena, was a dirty flying iceberg. This model was a radical departure from the then current model of a cometary nucleus as a collection of rocky particles—a flying dust cloud. To explain the enormous amounts of observed gas surrounding active comets, Whipple noted that for equal masses, ices are a much better source of gas than are rocks or dust. In addition, a collection of small rocky particles or dust would be quickly vaporized during a very close approach to the Sun. Surprisingly, a dirty iceberg would survive a close approach to the Sun better than a collection of rocky particles. Ices are generally poor conductors of heat so that only a surface layer of ice approximately one meter in depth is lost with each passage by the Sun. As the icy comet nucleus approaches the Sun, the solar radiation vaporizes the ices directly into gases, and the dust particles that are embedded in the ices are released. Whipple suggested that the primary ices of a cometary nucleus are water, methane, ammonia, and carbon dioxide. Much like a terrestrial iceberg, a comet's nucleus may not be very spherical in shape because its self-gravity during formation would not have been sufficient to produce the familiar globelike shapes of the larger bodies in our solar system. A typical comet's gravity is so small that

An artist's conception of the surface of a cometary nucleus. In response to solar energy, surface ices escape into space around the comet and drag dust particles along (from an original painting by William Hartmann).

the velocity required to escape the comet completely is only 1 or 2 meters per second (approximately 4 miles per hour). For comparison, to escape from Earth requires the thrust of an enormous rocket to attain a velocity of 11.2 kilometers per second (7 miles per second). A person standing on a comet, however, could "blast off" and achieve escape velocity with a good jump.

even to the edge of the solar system. By 1948, the speed of Comet Halley had fallen to 2000 miles (3200 kilometers) per hour, but its progress outward from the Sun was slower still, and declining: one mile per hour, one foot per hour, one inch per hour, then nothing. It was the turning point. Halley's Comet was 3¼ billion miles (5¼ billion kilometers) from the Sun.

The Sun glimmered in the distance, no longer the great glowing disk that lights and warms the Earth, but now showing itself for what it is: a star.

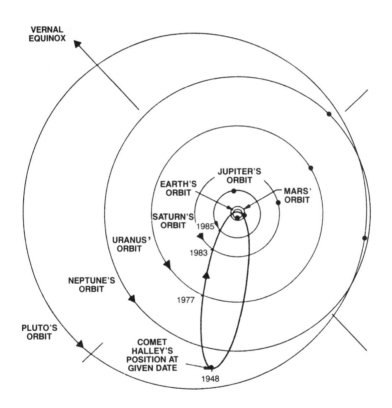

The orbit of Halley's Comet carries it closer to the Sun than Venus, then out beyond Neptune. Note that Comet Halley travels in the opposite direction of the planets (IHW diagram from The Comet Halley Handbook, *NASA).*

FALLING INWARD

The gravity of the Sun continued its pull, and the comet continued to respond: one inch, one foot, one mile per hour. The inward fall began again in 1948. A new generation would see the comet.

Just after Halley's Comet turned sunward, astronomers on Earth made important breakthroughs in understanding comets.

Until 1950, comets provided public spectacles and curious examples of orbital motions, but exactly what was a comet? The answer came from American astronomer Fred L. Whipple.

A comet is a dirty snowball. The nucleus of a comet is mostly water ice with tiny dust-sized flecks of rock. Whipple's concept of a comet explained and continues to explain better than any other model what we actually observe about the way a comet behaves. The previous hypothesis that a comet was a loose collection of gas and solid particles (a "flying sandbank") was rather quickly abandoned.

By 1950, Halley's Comet had turned inbound. Its path had carried it beyond the orbit of Neptune and also "southward," 928 million miles (1.49 billion kilometers) below the plane of the solar system, farther than Saturn lies from the Sun.[2]

Even at its farthermost point from the Sun, Comet Halley never reached the distance of Pluto's orbit, where the temperature stays close to –400 degrees Fahrenheit (–240 degrees Celsius). Pluto is the smallest of the planets, only 1900 miles (3000 kilometers) in diameter, smaller even than the moon of Earth. Its gravity is too weak to hold a gaseous atmosphere, yet Pluto manages to hold onto a moon of its own. This 750-mile (1200-kilometer) wide satellite was discovered in 1979 and named Charon (pronounced like the girl's name "Karen") after the boatman in Greek mythology who ferried the dead across the River Styx into Pluto's realm of Hades. Pluto and Charon travel a path around the Sun that carries them inside the orbit of Neptune, where they have been since 1979 and will be until the year 2000. Thus, for about 21 years out of every 248, Pluto is not the farthermost planet from the Sun. Because of its orbit, some astronomers think that Pluto and Charon were once satellites of Neptune. The composition of Pluto and Charon, probably mostly water ice, is very similar to a comet.

THE COMETARY DUST TAIL

As first suggested by Johannes Kepler in 1608, sunlight will exert a force upon cometary matter to form a tail in a direction opposite to that of the Sun. For dust particles smaller than one-thousandth of one millimeter, the pressure of sunlight is greater than the gravitational attraction of the Sun, so sunlight actually pushes the dust out of the coma and away from the Sun into the comet's dusty tail. The pressure exerted by sunlight is extremely low, but the dust is very small as well. Most of the solar radiation is emitted at a wavelength of approximately 4600 angstroms, and the radiation pressure is most effective on dust particles of comparable size. To the naked eye, dust tails appear yellowish-white because their light is the reflection of sunlight off countless dust particles.

Although many dust particles exist in a typical dust tail, the density is still very low. Even dusty comets such as Halley are not exactly choking with quantities of dust. When compared to the dust content of Comet Halley's tail, even the most immaculate household would be considered filthy. One million kilometers down Halley's tail, only one dust particle will be found for every houseful of space.

Dust tails usually develop to their maximum extent after perihelion because the most dust is released near perihelion, and it takes time for the solar radiation pressure to transport the dust into the tail region. In addition, the sharp curvature of the comet's path as it rounds the Sun broadens the tail in extent similar to the broadening of a stream of water from a garden hose when the hose is moved rapidly. Although dust tails are rarely seen when comets are beyond two astronomical units from the Sun, some comets such as Haro-Chavira (1956 I) and Bowell (1980b) have been observed out to four and five astronomical units with dust tails.

Dust tails can reach lengths of at least 100 million kilometers, but they are primarily confined to the plane of the comet's orbit. Dust particles that have been dislodged from the nucleus by the vaporization of the ice around them, but are too large to be easily pushed away from the comet by radiation pressure, remain concentrated around the nucleus or are gradually left behind along the comet's path. When a dusty comet is viewed along its orbit plane, we can sometimes see this concentration of large particles as an "antitail" seemingly pointing toward the Sun. Zdenek Sekanina of the Jet Propulsion Laboratory has studied the antitails of comets

Comet Arend–Roland (1957 III) displayed a normal tail flowing away from the Sun and an "antitail" apparently directed toward the Sun as it passed through the Earth's orbital plane in late April 1957 (photograph taken April 24, 1957, courtesy of the University of Michigan).

and concluded that this antitail phenomenon is due to sunlight being reflected from millimeter-sized dust particles. Furthermore, old short-period comets are likely to exhibit fainter antitails because their lower outgassing rates are incapable of lifting many of the larger dust particles off the surface of the nucleus.

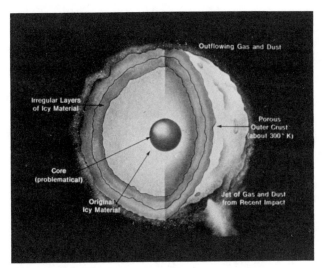

Most comet nuclei are probably not quite spherical. Near the Sun, a comet's surface is always losing materials to solar energy and the impacts of meteoroids. Whether comets have a core different from the rest of the "dirty snowball" is uncertain (reproduced with permission from Sky Publishing Company).

Ahead and far above Comet Halley's orbit lay Neptune, then Uranus—cloud-covered worlds, nearly twins in size. They too are enormous iceballs, with iceball moons—and with ice crystal rings, at least encircling Uranus. On January 24, 1986, as Halley's Comet approaches its closest point to the Sun, NASA's Voyager 2 spaceprobe will pass by Uranus at a distance of approximately 20,000 miles (32,000 kilometers) as it continues on a journey that carried it past Jupiter and Saturn and will, with a gravitational deflection from Uranus, hurl it toward Neptune for an encounter on August 24, 1989, and then out of the solar system forever.

Except for their monstrous size, Uranus and Neptune are so like comets in chemical composition that perhaps they are overgrown comets.[3] Perhaps Uranus, Neptune, and the comets formed together in this part of the solar system 4.6 billion years ago.

Comet Halley continued inbound and continued gathering speed. In 1982, while it was still beyond Saturn, master of thousands of rings and at least seventeen moons, the comet was detected by large telescopes on Earth, fainter than any comet that had ever been seen and no brighter than a candle viewed from 27,000 miles (44,000 kilometers) away. It would still be $3\frac{1}{3}$ years until Comet Halley would keep its 1986 appointment with the Sun. Yet already changes were occurring at the comet's surface. Perhaps the most volatile ices were beginning to vaporize, surrounding the solid nucleus with a coma—a shroud of gases and dust steadily expanding into space and forever lost to a comet because of the comet's weak gravity.

Onward and sunward was Jupiter, the largest of the planets, and, to a comet, the most dangerous. Each time Halley's Comet or any comet passed this way, the vast gravity of Jupiter and, to a much lesser extent, the gravity of Saturn, Uranus, and Neptune would accelerate or retard the comet's speed and bend its course so that its path would always be at least slightly changed. The proximity of Comet Halley's passages by Jupiter has varied the time required for the comet to complete one orbit from a period as short as 74.42 years (1835–1910) to as long as 79.25 years (A.D. 451–530). If a comet passes too close to Jupiter, the planet's gravity will hurl it out of the solar system, or divert it onto a new, shorter orbit around the Sun, or simply pull it apart.

Between Jupiter and Mars lay the asteroids, a belt of minor planets that formed with the solar system but never coalesced into a larger body because of the gravitational interference of Jupiter. These thousands of planetoids and billions of bits of rocky debris have a combined mass equal to approximately four percent of the mass of our Moon.

Here, at the distance of the asteroid belt, the heat of the Sun could begin to vaporize the less volatile water ice at the comet's surface. Comet Halley's coma, the cloud of gases expanding away from the comet's nucleus, was developing more rapidly. Comet Halley's orbit now carried it above (north of) the plane of the solar system.

Then the comet soared by Mars, a cold desert world whose rocky red surface is visible through its thin atmosphere. The

THE ION TAIL OF A COMET

In 1951, the German astronomer Ludwig Biermann found the outward motion of molecules in the tails of comets to be too large to be explained by radiation pressure alone. He suggested that a continuous outflow of ionized material from the outer atmosphere of the Sun blows the cometary ions back into the comet's tail. Later, satellite observations confirmed that indeed a "solar wind" does exist, just as Biermann had suggested.

The extreme temperature of the Sun's outer atmosphere (corona) creates a strong pressure that continuously carries away protons and electrons from the Sun's surface. In spite of the Sun's gravity, these charged particles rapidly accelerate away from the Sun. The Sun's magnetic field is "frozen" into this solar wind stream, and when this rapidly moving magnetic field encounters the much slower charged particles of the comet, the comet ions are trapped and dragged along "downwind" away from the Sun. Just as smoke pouring out of a truck's exhaust pipe is deflected in proportion to the relative velocities of both the truck and the outside wind, the deflection of a comet's ion tail depends on how fast the comet is moving and how fast the solar wind is blowing. At the Earth's distance from the Sun, the solar wind is traveling at approximately 400 kilometers (250 miles) per second. This solar wind is actually an extraordinary gale and is so swift that it completely controls the pokey cometary ions.

The solar wind blows the cometary ions near the nucleus straight back into the ion tail. Because the ions spiral around the solar magnetic field lines, the tail ions follow the straight and narrow path. The ion tail extends nearly straight outward from the comet nucleus in a direction opposite that of the Sun.

The cometary ion most readily seen in ion tails is carbon monoxide that has been stripped of one electron (CO^+). Radiating primarily near 4500 angstroms, this ion appears blue to our naked eye. Other ions observed in tails include positively charged fragments of water, methane, and ammonia: H_2O^+, OH^+, CH^+, and N_2^+. Although most ion tails do not form when the comet is beyond two astronomical units from the Sun, exceptions do occur. Comet Humason (1962 VIII) showed strong CO^+ spectra out to five astronomical units.

The ion tail of short-period Comet Giacobini–Zinner as seen on October 26, 1959 (official U.S. Naval Observatory photograph taken by E. Roemer).

surface of Mars shows numerous craters where asteroids and, undoubtedly, comets also have struck over the past 4.6 billion years.

The journey sunward continued and the speed of the comet increased. The comet warmed as it approached the Sun, and its coma grew. Beyond the coma, a very tenuous cloud of hydrogen expanded.[4] The pressure of sunlight and fast-moving particles from the Sun pushed some of the dust and gas out of the coma, and the comet began to develop a tail.

A QUICK STOP AT EARTH?

While inbound on this journey to the Sun, Halley's Comet will come no closer to Earth than 57.6 million miles (92.8 million kilometers). This passage occurs on November 27, 1985. Outbound from the Sun, on April 10, 1986, the comet will be 39.0 million miles (62.8 million kilometers) from Earth, its closest approach this circuit. But Comet Halley has swept by much closer in years past. In A.D. 837 it passed 3 million miles (4.9 million kilometers) from Earth. Other comets have come still closer.

The modest gravity of Earth has changed the flight of close-passing comets. But comets in turn have never measurably affected

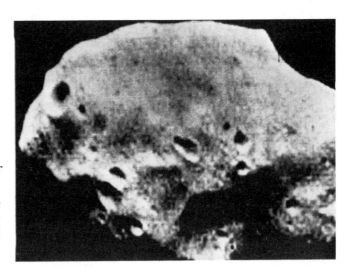

Phobos is the larger of the two moons of Mars. Both moons appear to be similar to asteroids (Mariner 9 photograph courtesy of NASA).

the motion of Earth. Because comets are so easily deflected by gravity yet do not disturb the planets or even their moons, astronomers have calculated that an average comet's nucleus must weigh less than one-trillionth of the Earth. The nucleus of a typical comet (such as Halley) must be only approximately 1 to 5 miles (roughly 1 to 10 kilometers) in diameter.

Comets and asteroids may be small in mass, but they are very numerous. And, traveling among the planets, they constitute a hazard. Because the solar system formed 4.6 billion years ago, the Earth, Moon, and planets have been moving targets for billions of asteroids and comets, as the cratered surfaces of our Moon, Mars, and Mercury testify.[5] The Earth has been hit even more often than our Moon. It is a larger target. But the erosive forces of wind, water, and geological activity steadily erase the craters.

Yet other evidence remains. Possibly, in two different ways, comet impacts have been indispensable to the rise of life and, ultimately, human life on Earth.

THE CARBON DUMP

Astrochemists, analyzing the light from gases in comet heads and tails, have already detected in comets carbon, oxygen, hydrogen, nitrogen, and sulfur—all the chemical elements necessary for life except phosphorus. Comets crashing into our newly formed planet must have contributed significantly to the supply of organic chemicals from which life on Earth arose.

John Oró estimates that comet impacts in the early days of Earth contributed 100 million billion tons (10^{23} grams) of the organic elements needed for life, including 10 million billion tons of carbon, the element that is the foundation of all life. This organic chemical abundance does not imply that comets deposited living organisms on Earth. These chemicals may have entered our atmosphere as complex organic molecules, including amino acids, the building blocks of proteins, which, in turn, are the building blocks of living organisms. Such molecules have been found in certain interstellar gas clouds and some meteorites, but the heat of impact probably broke up the complex molecules. The organic elements, however, were still intact, and the Earth was a friendly environment, made even more hospitable, ironically, by the violent comet impacts that provided much of the water and other gases to the Earth's atmosphere, perhaps even more than the gaseous outpourings of volcanoes.

Much of the Earth's original carbon must have been buried out of reach below the surface or bound up chemically in rocks. Comet crashes provided more carbon, oxygen, hydrogen, nitrogen, and sulfur than was accessible before. The same chemical reactions that occurred in nebulas, on asteroids, and on comets began anew and continued to develop in complexity. Soon, perhaps only one-half billion years after the Earth formed, life appeared.

Increasing evidence suggests that comets were the chemical engineers that enhanced the chemical supplies and adjusted the terrestrial environment for the origin of life.[6]

GONE WITH THE COMETS?

Growing evidence indicates that comets may have played decisive roles at major turning points throughout Earth history. Sixty-five million years ago, for example, a great number of dinosaurs, the largest animals on Earth at that time, suddenly became extinct. At that same time, more than one-half of all animal and plant species also perished.[7]

It was not the first time the Earth had lost much of its life. Almost all the species of plants and animals that have ever lived on our planet are now extinct. The current fossil record documents the disappearance of some 200,000 different forms of life.[8]

Over the past two centuries, geologists and paleontologists have steadily delineated layers in the Earth's crust on the basis of the remains of plant and animal species found in one stratum but not in adjacent ones. Growing evidence suggests that periodically a substantial percentage of life forms on Earth have become extinct.

But what caused these mass extinctions? Were they attributable to changes in climate, geological activity, fluctuations in ocean levels, or sudden influxes of cosmic rays? No answer or combination of answers was satisfactory.

Then, in 1980, a new explanation appeared.[9] Few scientific papers in recent times have generated more excitement. The authors were Nobel Prize-winning physicist Luis W. Alvarez, geologist Walter Alvarez (his son), and nuclear chemists Frank Asaro and Helen V. Michel. They had noticed that between the Cretaceous and Tertiary strata of 65 million years ago was a thin layer of clay containing an excessive amount of the rare metal iridium. Not much iridium exists in the Earth's crust because soon after our planet formed 4.6 billion years ago, it became molten because of the heat developed in its interior by gravitational pressure and because of the heat-producing decay of radioactive elements scattered throughout it. In this early molten phase, the heavy elements—iron, nickel, iridium, and others—sank toward the center to become the Earth's dense core. The lighter materials floated to the surface. The Earth "differentiated." Consequently, scientists interpret unusual amounts of iridium on the Earth's surface as a contribution from celestial objects: meteoritic particles from asteroids and comets.

But why should asteroids and comets have more iridium than the Earth? They probably don't, averaged as a whole. The cloud of gas and dust from which the Sun, planets, moons, asteroids, and comets formed probably had a fairly homogeneous composition. But the comets and most of the asteroids were small bodies. They were heated somewhat by the radioactive elements they contained, but they had too little mass and therefore too little gravitational pressure to become molten. Their iridium and other heavy elements never sank to their centers. So, fragments of the asteroids and comets (or even entire comets) contain a much higher fraction of iridium than the surface of the Earth.

When this Cretaceous–Tertiary boundary layer at sites in Italy, Denmark, and New Zealand exhibited extraordinary amounts of iridium, it meant that a meteorite or comet had been there. Furthermore, the object must have been large because its debris was scattered all over the planet. In fact, its size could be estimated by calculating the amount of iridium in the boundary layer worldwide and then figuring how large a meteorite or comet would be required to deposit that much iridium. The celestial killer was calculated to have been 6 miles (10 kilometers) in diameter.

What made that impact so different from most of the trillions of meteorite falls that preceded it was its size and the fact that its collision coincided with the extinction of the dinosaurs and at least

THE ROCKET EFFECT
OF A COMETARY NUCLEUS

While working on the orbital motion of the comet that bears his name, Johann Encke (1791–1865) noted that his comet's orbit was shrinking with time. During the nineteenth century, Comet Encke consistently arrived at perihelion approximately 2.5 hours ahead of the prediction. In the early twentieth century, orbit computers noted that the orbits of certain other comets, such as short-period comets Wolf and d'Arrest, were not shrinking but rather expanding with time; these comets were arriving at perihelion slightly after the predictions. This state of affairs was unsettling and was particularly strange because the same techniques that did not seem to work too accurately for predicting the motion of comets worked with great precision when applied to planets and asteroids. Apparently some

Schematic diagram showing the possible effects of nongravitational forces on the motions of comets.

fifty percent of all the plant and animal species on Earth at that time. Its impact, the research team suggested, had thrown upward one thousand times as much dust as a large volcanic explosion. The sky was blackened. Winds spread the dark cloud of death around the world. Day became as black as night for months.

additional force that had not been included in the orbital predictions was affecting the motions of some comets, and this force had nothing to do with gravity.

According to the dirty iceball cometary model suggested by Fred Whipple, the anomalous motion of some comets was due to a rocket effect caused by gases vaporizing from the cometary nucleus. In fact, one of the motivating reasons Whipple put forward his dirty iceball hypothesis was to explain these nongravitational accelerations in comets. According to Whipple's hypothesis, solar radiation falling on a globe of ice will cause most of the ice vaporization to occur in the sunward direction (*see* a in the schematic). The gases streaming off in a sunward direction push the comet slightly outward, away from the Sun, just as Newton had shown: For every action, there is an equal and opposite reaction. However, if the nucleus of the comet is rotating, the time lag between the maximum solar radiation received at cometary noon and the time when the heating and vaporization are greatest will cause the thrust to be directed a little bit forward or backward and thereby will alter the comet's orbit and motion slightly. Depending on whether the cometary nucleus is rotating in a direct or retrograde fashion, the comet's orbital velocity will either be increased or slowed, and the orbit will consequently expand or shrink in size (*see* b and c in the schematic). For some comets, this rocketlike thrusting of the nucleus can cause appreciable effects on the comet's motion. For Comet Halley, this nongravitational force causes the comet to arrive at perihelion four days later than it would if this effect were inoperative. Hence, the nucleus for Comet Halley must be rotating in a direct sense, that is, in the same direction as the comet's orbital motion (as in c in the schematic).

Plants, starved for sunlight, died. Plant-eating animals, starved for plants, died. Meat-eating animals, starved for other animals, died.[10]

Gradually the dust fell back to Earth as a thin layer of clay, leaving a clue to what had befallen the dinosaurs.

Additional studies showing high levels of iridium and other similarly rare (hence mostly of extraterrestrial origin) elements at more than fifty sites throughout the world have continued to strengthen the view that the dinosaurs died of cosmic causes.

A COSMIC RHYTHM

In 1984 came more surprises. Paleontologists David M. Raup and J. John Sepkoski, Jr. examined the dates Sepkoski had painstakingly determined for the extinction of 567 different families of marine creatures and found that mass extinctions occurred at intervals of approximately 26 million years.[11] Their mathematical analysis enabled them to assign a very high probability (more than ninety-nine percent certain) that these extinctions constituted a real cycle and were not just random patterns. Within the past 250 million years, at least eight and perhaps as many as ten mass extinctions had occurred.

Walter Alvarez and Richard A. Muller followed with a review of the dates when meteorite craters with diameters exceeding 6 miles (10 kilometers) were formed, based on research by Richard A. F. Grieve.[12] They found that these large craters had been gouged out at intervals of 28.4 million years, essentially the same cycle (within the margins of dating uncertainty) as the mass extinctions. In addition, the dates of the large craters corresponded almost exactly with the dates of the mass extinctions. The two cycles were in phase.

Prior to 1984, neither mass extinctions nor large-scale cratering was suspected of showing cycles. Now they were. Mass extinctions and large impact craters were not considered to be related to one another. Now they were. Excessive iridium was found in the mass extinction boundary layers of three other epochs. It looked as if the process of forming large craters made a hole in life on Earth as well.

But what caused the craters? What gave the Earth a beating every 28.4 million years?

Sizeable asteroid fragments strike the Earth occasionally, but not on a regular rhythmic basis. Therefore, attention shifted from asteroids to comets. The paths of comets had convinced astronomers since 1950 that a spherical cloud of billions or trillions of comets surrounds the fringe of our solar system like a thick hollow ball. A few of these comets plunge into the realm of the planets whenever a star or dense cloud of interstellar gas passes close enough to our solar system for its gravity to disturb some of these icy sentinels.

The impact of a large comet, whether it smashed into the ground, splashed into water, or broke up in the air, would give rise to the same dark Sun-obscuring cloud. Not only is the mass of a comet approximately half dust and half ice, but a large comet that struck an ocean would still form a crater on the sea bottom and throw vast quantities of dust into the atmosphere. And comets can hit much harder than asteroids despite their fragility and smaller average size. The reason is that all asteroids revolve around the Sun in the same direction as Earth. When they reach the Earth's orbit, they are traveling approximately the same speed as Earth and in pretty much the same direction. It is like driving on a highway and colliding with a car traveling in the same direction. We must overtake it or it must overtake us. Either way, the difference in velocity is rather low. In the case of asteroid collisions, the speed at impact is never much higher than 56,000 miles (90,000 kilometers) per hour. However, such a collision produces plenty of damage. For example, the Barringer Meteor Crater near Winslow, Arizona, despite 25,000 years of erosion, is still four-fifths of one mile (1^1/$_3$ kilometers) in diameter and 600 feet (180 meters) deep. The crater is the bullet wound burrowed into the Earth by an iron meteorite only about 90 feet (28 meters) in diameter.[13]

Comets can hit harder than asteroids because approximately one-half of the comets revolve around the Sun in retrograde orbits—the opposite direction of Earth. Colliding with them is like zipping down the highway and ramming head on into a car coming from the opposite direction. In the case of comets with retrograde orbits, the speed of impact can be as high as 160,000 miles (260,000 kilometers) per hour.

But what could disturb the comet cloud approximately every 26 to 28 million years and launch a deluge of icy missiles toward the inner solar system to cause a major extermination of life on Earth?

Two proposals that captured much scientific interest appeared immediately.

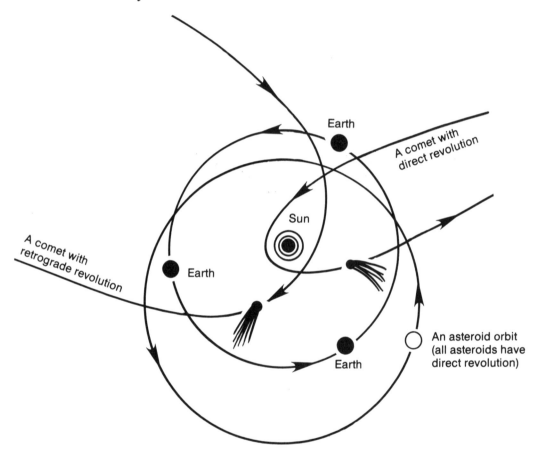

Typical asteroid and comet orbits demonstrating that comets that collide with the Earth are most likely to be traveling much faster than asteroids.

AN UNSEEN COMPANION

Two independent teams of researchers concluded that a star bound by gravity to our Sun might be the instigator of the repeated mass murders of life on Earth.[14] We had never realized it before, they said, but our Sun might be part of a binary star system. And why not? A majority of the observable stars have one or more companion stars orbiting with them. Alpha Centauri, the nearest known star to the Sun, is a three-star system. Our Sun, they said, might have a previously undetected companion that travels out to a distance of approximately 2.4 light years and then returns to a distance of approximately one-half of a light year—a distance within the cloud of comets and with sufficient gravity to change the course of billions of comets.

But how could we have missed seeing the second most massive and second most luminous object in our solar system? The answer was that we are now almost midway between comet showers, so if the binary companion exists, it must be close to its maximum distance of 2.4 light years from us. Our Sun's companion would be a small faint red star with perhaps one-tenth the mass of our Sun and a probable brightness of magnitude 12, more than 200 times too faint to be seen with the unaided eye. It could have sent billions of icy assassins descending upon Earth life without revealing its position in the sky. But its image would appear among the million brightest stars recorded in photographic atlases made by telescopes. It might be recognized by its parallax, its tiny but significant shift in position against the more distant background stars when seen from opposite sides of the Earth's orbit.

If the Sun does have an abusive twin, one of the teams of scientists—Marc Davis, Piet Hut, and Richard A. Muller—suggested that "it be named Nemesis, after the Greek goddess who relentlessly persecutes the excessively rich, proud, and powerful." Then they added, "We worry that if the companion is not found, this paper will be our nemesis."

Paleontologist Stephen Jay Gould offered a different name: Siva, the Hindu god of death. In addition to being more ecumenical, Gould felt that Siva would be more accurate because

COMETARY EXPLOSIONS

Many cometary observers have been surprised by the sudden brightening of a comet that gave no telltale signs of aberrant behavior. On the average, comets that flare will show a brightness increase of 100 times their former luster, but brightness increases of 1000 times are not uncommon. Cometary brightness flares are completely unpredictable. New comets fresh from the Oort Cloud have shown brightness flares, but so have the much older short-period comets. No preferred orbital location or solar proximity has been exhibited by the comets that have displayed brightness flares. A comet is just as likely to flare at a great distance from the Sun as in the solar neighborhood.

Spectra taken immediately after several flare events usually reveal a continuous spectrum, which indicates that sunlight is being reflected off cometary dust particles. Thus, the flare event is probably caused by a sudden increase of dust particles surrounding the comet. Scientists are not sure what triggers the flare event. It may be the explosive release of a pocket of volatile gas just exposed on the comet's surface, or it could be the collision of a cometary satellite or interplanetary boulder with the comet.

Many comets are discovered during flare events or during periods of abnormal brightness. These brightness flares appear to be most common for those comets that are arriving from the Oort Cloud for the first time. These so-called new or young comets may have a thin layer of volatile ices that fluoresce at large heliocentric distances on their way into the solar neighborhood. On their way out of the solar neighborhood and on subsequent returns, this extra-volatile layer is gone so that a new comet's first passage near the Sun is often its most impressive. This extra-volatile layer also may explain why so many new comets appear intrinsically brighter before perihelion than they do after. The disappointing postperihelion performance of Comet Kohoutek (1973 XII) may have been due to this effect.

Siva does not punish; he simply terminates life so that other life may go forward.[15]

A serious problem with the "death star" proposal is the instability of the companion star's orbit. The passage of our binary star system close to other stars or near a relatively dense cloud of interstellar gas would gravitationally disturb the motion of our Sun's small twin and thus disrupt the regularity of its period and soon tear it loose from the Sun's gravitational grasp altogether.

THE GALACTIC CARROUSEL

An alternative proposal was offered simultaneously by Michael R. Rampino and Richard B. Stothers.[16] They blamed no stellar twin, but instead the motion of the solar system within the Milky Way. Our Sun is one of approximately 400 billion stars in our galaxy. We lie some 30,000 light years from the center, three-fifths of the way toward the edge. If we could look at our galaxy from the edge, we would see the shape of a fried egg. Seen from the top, our galaxy is a spiral. Our Milky Way is, in fact, a normal but larger-than-average spiral galaxy. Like the planets in orbit around the Sun, our Sun is in orbit around the center of the Milky Way. Even at 145 miles (235 kilometers) per second, it will take our Sun about 230 million years to complete one circuit of the galaxy.

Our Sun and its planets are situated in the spiral arm region of the Milky Way and lie within the disk, more or less. Not only is our solar system speeding along in its orbit, but it is also oscillating slightly up and down—in and out of the plane of the Milky Way, a little like a dolphin swimming forward and "porpoising" up above and then down below the surface of the water. The solar system, of course, does not porpoise purposely. Instead, it was probably formed just a bit above or below the exact plane of the galaxy so that the combined gravity of the nearby stars in the galactic disk pulled the solar system into the disk. When our solar system reached the midplane of the galaxy—its vertical motion at a maximum—it overshot and slid toward the other side of the disk. The combined gravity of the disk stars then slowed the Sun's vertical motion to a stop, and it fell back the other way through the

When a spiral galaxy like our Milky Way is seen edge on (above), it appears rather flat. Seen face on (next page), it shows its spiral form. Our solar system lies in the galactic disk three-fifths of the way out from the center of the Milky Way (left photograph courtesy of Mount Wilson and Las Campanas Observatories of the Carnegie Institution of Washington; right photograph courtesy of Palomar Observatory).

plane of the Milky Way, over and over again. For our Sun, the cycle from one midplane passage to the next is estimated to require approximately 33 million years. As the solar system dives through the galactic plane, it also plunges past the greatest concentration of dense clouds of interstellar gas and dust in the galaxy. The gravity of this interstellar material would disturb the motion of the planets very little because of their large masses, but it could alter significantly the motion of the small outlying comets. Some would fall into the inner solar system over a period of approximately 1 million years, and the resulting comet bombardment would have the same devastating effect as the close approach of a binary star companion to the Sun's comet cloud.

One major objection to the nebula passage hypothesis is that our solar system is now near the galactic plane, so we should be near the peak of a comet bombardment, but the cratering and extinction records say this peak happened approximately 11 million years ago, when we were almost a maximum distance from the midplane of the galaxy. Also, the 33-million-year porpoising cycle does not correspond very closely to the currently measured extinction or cratering cycles.

IMPLICATIONS FOR EVOLUTION

Still, a gravitational perturbation caused by a companion star of our Sun or by an interstellar gas cloud could send several billion

A COMET DUNIT?
(TUNGUSKA, SIBERIA, 1908)

On the morning of June 30, 1908, a brilliant fireball visible in broad daylight swept across the skies of central Siberia. It raced from southeast to northwest, leaving a train of flame and smoke behind it. At 7:14 A.M., the trail ended 600 miles (1000 kilometers) north of Irkutsk near the Stony Tunguska River in a thunderous explosion heard over 600 miles away.

The concussion hurled a man off his porch at the trading post of Vanavara 40 miles (65 kilometers) away. The shock waves leveled

Ilya Potapovich Petrov (previous page, left) lived close to the Tunguska blast area and gave accounts of destruction that included the death of a large reindeer herd. S. B. Semenov (previous page, right), a farmer living sixty-five kilometers from the site, had his shirt nearly burned off his body and was blown off his porch during the Tunguska Event on June 30, 1908. The photograph above, taken nineteen years after the blast, shows the charred and fallen trees near the blast site (photographs courtesy of E. L. Krinov, Soviet Academy of Sciences).

a vast forest, toppling trees outward from the blast to a distance of 18 miles (30 kilometers).

A wave of heat that preceded the shock wave incinerated most of a herd of 1500 reindeer and scorched the forest trees for 11 miles (18 kilometers) in all directions.

The total blast damage done was equal to a large hydrogen bomb. Yet, fortunately and astonishingly, no human being was killed in the Tunguska Event.

A COMET DUNIT?—*cont.*

Tiny particles of dust from the shattered object and from the ground were sucked into the upper atmosphere and spread by (the then unknown) jet stream winds around the Northern Hemisphere where for several nights they caused unusually bright skies and auroralike displays and, for months, eerie red sunsets and sunrises.

What had happened?

The Tunguska region was so remote (and Russian affairs through this period so chaotic) that no scientific team reached the devastated area until 1927, nineteen years after the event.

Leonid Kulik, head of that first modest expedition, expected to find a large crater where a meteorite had struck, but no crater was found. Instead, embedded in the tree trunks, standing and fallen, and in the soil were many tiny flecks of metal and beads of glass. These grains and globules, averaging approximately one-tenth of one millimeter (four-thousandths of one inch) in size, had the chemical composition not of earthly particles but of asteroids and comets.

Clearly, the object that created the fireball came from space and had disintegrated in the air at an altitude of approximately 5 miles (8 kilometers), scattering particles for miles. No large pieces had hit the ground; hence, no craters and no great meteorites were found.

In 1930, English meteorologist Francis J. W. Whipple (no relation to American astronomer Fred L. Whipple) suggested that the celestial intruder was a small comet rather than a rocky or metallic fragment of an asteroid.

For half a century this explanation seemed satisfactory. No crater was found because a comet is mostly ice, and the friction of its blazing passage through the air turned most of its mass into gas. The tiny metallic and glassy particles found in the trees and soil could be the dust of the "dirty snowball" fused by the heat of its collision with the Earth's atmosphere.

In 1978, Czech astronomer Lubor Kresák proposed that the Tunguska Event was wrought by a specific comet, a fragment of Comet Encke, the comet with the shortest orbital period (3.3 years) of any known.

However, in 1983, a powerful analysis by astronomer Zdenek Sekanina rejected the comet explanation. Small intruders from space that hit the Earth's atmosphere slow down rapidly because of air friction. But the atmosphere is too rarefied to decelerate efficiently something as large as the Tunguska visitor, which would have covered a major league baseball field and had a mass of several million tons (10^{12} to 10^{13} grams).

The Tunguska object was too tough to crumble as it began its final fall. It held together down to the height of Mount Everest. The Earth overtook the Tunguska object and it entered our atmosphere traveling approximately 31,000 miles per hour (14 kilometers per second) relative to Earth. It was still going 22,000 miles per hour (almost 10 kilometers per second) when it exploded. It disintegrated, says Sekanina, because of excessive aerodynamic pressure. It reached a lower, denser region of the atmosphere at such a velocity that the effect was millions of times worse than a speeding car hitting a stone wall. The energy of the object's motion was converted to heat that vaporized most of the material. Therefore, no large chunk was left to fall to the ground, and only tiny heat-fused rock and metal fragments were scattered over the landscape. This single terminal breakup, says Sekanina, is more characteristic of rocky asteroids. A comet, being a more fragile mixture of ice and dust, would never have survived its atmospheric flight to such a low altitude.

Further, argued Sekanina, even at an atmospheric entry speed of 31,000 miles per hour, the interloper was traveling much slower than any comet ever observed passing close to Earth. And in any case, concluded Sekanina, the fireball's path in the sky did not fit the orbit of Encke's Comet.

So what was this brightest and most massive fireball on record? The culprit, Sekanina said, was a meteoroid after all—an Apollo object, a special class of asteroids with orbits that cross that of the Earth. The speeds and paths of these rocky bodies fit quite nicely the trajectory of the Tunguska fireball.

But there may yet be a "comet connection." Ernst J. Öpik and George W. Wetherill have suggested that almost all Apollo objects are really just ice-exhausted, gaseously inactive nuclei of comets.

The controversy continues as to whether the Tunguska blast was caused by an asteroid or a comet.

Sekanina says we can expect to get hit by a Tunguska-size meteoroid every 2,000 to 12,000 years.

*The hazy band of the Milky Way marks the plane
of our galaxy seen from within. The bright and
dark clouds of gas and dust within our galaxy
might disturb the outlying comet cloud of our Sun
as it passes through the galactic plane (reproduced
with permission from Hans Vehrenberg).*

comets tumbling toward us. Instead of about five new comets reaching the inner solar system each year, we would expect 5000 to 10,000—one per hour! This celestial fusillade is too great for the Earth to escape injury. From 10 to 200 comets would be expected to hit our planet over a period of 1 million years—a thousand times the normal impact rate.[17]

About 250 million years ago, the Permian Period ended abruptly as ninety-six percent of its creatures perished. Thirty-eight million years ago, the Eocene Period terminated suddenly with a loss of fifty percent of its species.

Rhythmic mass extinction due to cosmic causes has enormous implications for biological evolution. Led by Stephen Jay Gould and Niles Eldredge, scientists recently have recognized from the fossil record that evolution does not proceed by steady change. Instead, it proceeds by "punctuated equilibrium," in bursts followed by long periods of very little change. Celestial impacts might offer a mechanism to explain the largest of these bursts. Each mass extinction allowed the survivors a chance to fill the environmental niches left by the victims. Sixty-five million years ago, a comet may have eliminated the predator reptiles that held down the mammal population. In the mammal "explosion" that followed, one of the eventual evolutionary results was human beings.

It is fascinating, then, to consider that evolution proceeds far less by competition among life forms than by struggle against environmental disasters triggered by unearthly causes. When devastation comes, it tends to wipe out the strong and weak alike. Nowhere, says Gould, do we find in evolution the internally driven upward advance whereby we proudly see mankind as evolution's objective and crowning achievement.[18] We may be, instead, the accident between disasters.

Periodic mass extinctions likewise have intriguing implications as we search for intelligent life beyond the Earth. Presumably all stars are surrounded by comet clouds. If the encounter with a binary star or a nebula regularly sends a flight of comets on a mission of mass extermination, does this keep an evolutionary

Comet Halley and planet Venus.
May 13, 1910 (photograph
courtesy of Lowell Observatory)

system from stagnating and make possible the rise of intelligent life? On the other hand, how often does life simply not survive the onslaught? Or do intelligent civilizations arise but fail to last long enough to master interstellar travel and spread throughout the galaxy? Evolutionarily, we know of no reason why life elsewhere in our Milky Way could not be 1 billion years or more ahead of us. Yet where are they?

And next time? The last great extinction on Earth was approximately 11 million years ago. Will it be humanity's turn 15 million years from now?

JAWS OF DEFEAT

For many centuries, people had feared comets as portents or even agents of havoc and death. The work of Tycho, Newton, and Halley gradually changed that perception. Comets were fascinating, even pretty. Awards were given for their discovery. Careful study yielded information on comet paths, activity, and composition, which in turn shed light on the origin of the solar system and the development of life therein. These findings led to the conclusion that we may well have good reason to fear comets after all.

But Comet Halley is no threat, at least not for many centuries to come. It is not presently on a collision course with any of the inner planets and keeps far enough from the massive outer planets to avoid a major orbital change for thousands of years.

NEARING THE SUN

And so the cycle continues. Since it leisurely turned sunward in 1948, the pace of Comet Halley's appointment schedule has steadily quickened: the orbit of Neptune in May 1963, Uranus in March 1977, Saturn in April 1983, Jupiter in December 1984; then frenetically onward: the orbit of Mars in November 1985, Earth on New Year's Day 1986, and, still gathering speed, Venus, the last of its inbound orbital crossings, three weeks later.[19]

Did comet impacts, which brought significant quantities of life-giving chemicals to the surface and atmosphere of Earth, provide the same service to Venus? If so, it was too much of a good thing. Venus developed a thick atmosphere. But it lies closer to the Sun than Earth and therefore receives more solar energy. The carbon dioxide in the atmosphere prevents the excess solar heat from escaping, so Venus roasts at a temperature of 900 degrees Fahrenheit (480 degrees Celsius).

No doubt the same chemicals were deposited by comets onto the surface of Mercury as well. Just as no life arose on Venus, no life appeared on Mercury, but for a different reason. The gases contributed by comets floated away. Mercury has too little gravity to hold an atmosphere.

Fortunately, Comet Halley will not venture as close to the Sun as Mercury because some comets passing close to the Sun have been observed to break into pieces under the intense tidal strain of the Sun's gravity.[20]

BARBECUING A COMET

On February 9, 1986, Halley's Comet reaches its perihelion, its closest approach to the Sun: 54.5 million miles (87.8 million kilometers). It is traveling 34 miles (54.6 kilometers) per second, the equivalent of flying from New York to Los Angeles in just over one minute or halfway to the Moon in one hour.

The intense energy of the Sun prepares the comet for its best display. Near perihelion, gases, mostly water vapor, boil off the comet's icy nucleus at the rate of more than one million tons per day.[21] Even so, Halley's Comet has sufficient ices to endure at least one thousand passages by the Sun.

From the evaporating ices of the nucleus, a gas cloud, the coma, expands to the size of Jupiter. Beyond the coma, a hydrogen cloud having a diameter greater than the Sun develops. The comet's tail is already millions of miles long and still lengthening. It could reach 50 million miles (80 million kilometers) in length. Some comet tails grow to more than 100 million miles in length, greater than the distance between the Sun and Earth.

The dimensions of the coma, the hydrogen cloud, and the tail are large, and their display is memorable, but very little material is present. Stars shine through a comet's tail and all but the center of the coma with no loss in brightness. It has been said that a comet is the nearest thing to nothing anything can be and still be something.[22]

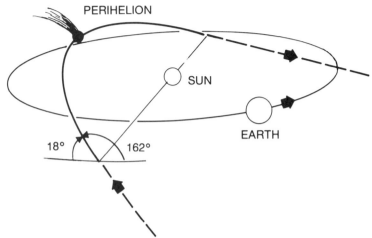

PERIHELION

SUN

EARTH

18° 162°

Comet Halley's orbit is tilted so that it cuts through the plane of the Earth's orbit. When close to the Sun, the comet is north of the Earth's orbit. When far from the Sun, the comet is well south of the planets (based on a diagram courtesy of the IHW).

As the comet rounds the Sun, two parts of the tail become apparent. One part extends outward from the Sun but is distinctly curved. This portion of the tail is caused by the pressure of sunlight. The pressure of sunlight pushes away tiny particles of dust. These particles, lost to the comet, are now in orbits of their own, orbits that are farther from the Sun; therefore, in accordance with the laws of planetary motion that Kepler described and Newton explained, these tiny particles now travel more slowly. They lag behind the comet nucleus and give this dust tail its curved appearance. The dust particles shine by reflecting the Sun's light, and so the dust tail is yellowish-white.

The other part of a comet's tail stands almost straight out from the coma, away from the Sun. Solar radiation imparts such energy to the coma gases that it causes them to become excited and to ionize, that is, to shed electrons, so that when the gases snare other electrons and return to a de-excited state, they glow by fluorescence, essentially by the same process as a neon sign. The coma ions are transported tailward by the solar wind—charged

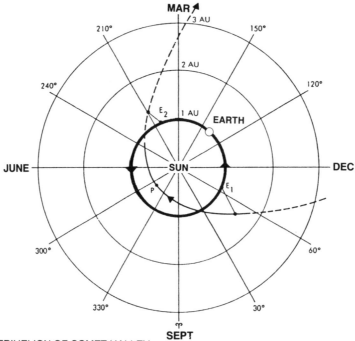

P = PERIHELION OF COMET HALLEY
E_1 = POSITION OF EARTH AT PRE-PERIHELION CLOSE APPROACH OF COMET (NOV. 27, 1985)
E_2 = POSITION OF EARTH AT POST-PERIHELION CLOSE APPROACH OF COMET (APR. 11, 1986)
O = POSITION OF EARTH AT PERIHELION OF COMET HALLEY (FEB. 9, 1986)

*Relative positions of Halley's Comet and Earth in 1985–86.
The orbit of Halley's Comet is shown as an unbroken line
when the comet is north of the Earth's orbit and as a broken
line when the comet is south of the Earth's orbit (IHW
diagram from* The Comet Halley Handbook, *NASA).*

subatomic particles from the Sun traveling at almost 1 million
miles (1.4 million kilometers) per hour and sometimes faster. These
tiny particles blast away at the gases in the comet's coma and carry
them rapidly away from the Sun. This stream is the ion tail. To our
eyes it appears bluish-white because of the fluorescence of ionized
carbon monoxide (CO^+).

The response of a comet's tail to the Sun makes it a natural
solar-wind sock. The comet's tail changes in response to variations

Comet West (1976 VI), March 4, 1976 (reproduced with permission from Hans Vehrenberg).

The impact of a comet or asteroid may have been responsible for the extinction of more than fifty percent of all plant and animal species on Earth 65 million years ago, including the dinosaurs (painting by Don Davis).

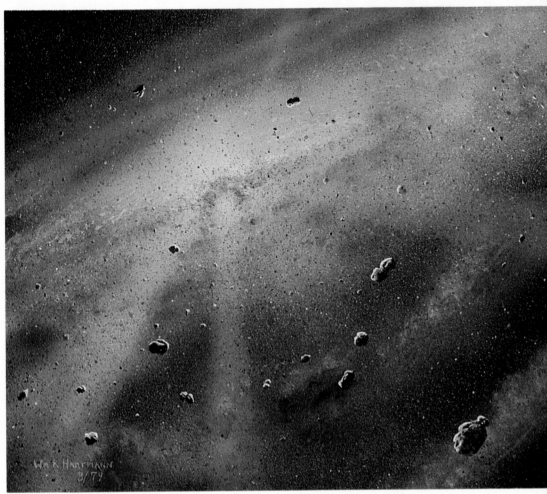

View of primordial solar nebula with newly formed planetesimals (from an original painting by William Hartmann).

Meteor crater in northern Arizona was caused by an iron meteorite, not a comet, but demonstrates that relatively small objects can cause much damage. The Earth has been and continues to be hit by objects from space (photograph courtesy of Meteor Crater, northern Arizona).

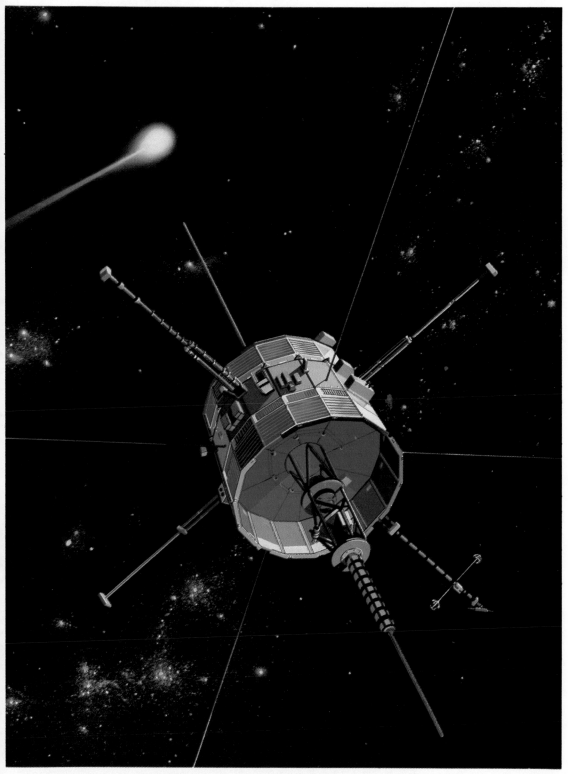

*An artist's conception of the International Cometary Explorer
(ICE) spacecraft approaching Comet Giacobini–Zinner in
September 1985 (courtesy of NASA Goddard Space Flight Center).*

The Bayeux Tapestry features a scene showing the English shrinking from Halley's Comet while King Harold hears of the bad omen from an adviser. The date was 1066. Later that year, the invading Normans won the Battle of Hastings. It is thought that Bishop Odo, William the Conqueror's half-brother, of Bayeux, France, commissioned this work about 1070. This reproduction of the original was created by Delphine Delsemme (photograph courtesy of Armand H. Delsemme).

Halley's Comet over England in 1682 when Halley himself saw it. This painting was made by David Teniers the Younger. From L'Illustration, *Vol. 135, May 7, 1910, p. 428 (photograph courtesy of Ruth Freitag).*

"The Adoration of the Magi" by Giotto di Bondone, painted in the early fourteenth century on the ceiling of the Scrovegni Chapel in Padua, Italy. It is thought that Giotto may have used the 1301 apparition of Comet Halley as a model for the star of Bethlehem (from the collection of D. K. Yeomans).

The Comet of 1577 as it appears in an Islamic manuscript that was made at the Istanbul Observatory in the sixteenth century. Courtesy M. Dizer, Director, Kandilli Observatory, Istanbul, Turkey (photograph courtesy of O. Gingerich).

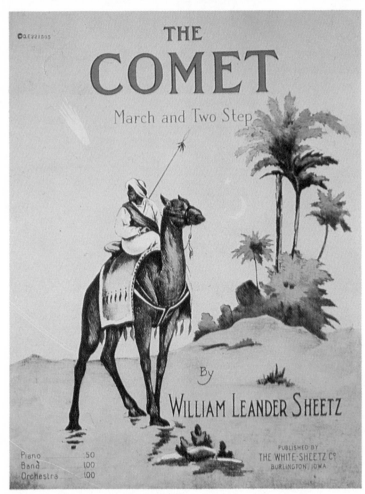

Sheet music cover from 1909.

1910 German postcards (set of five) (from the collection of D. K. Yeomans).

Left and right, sheet music covers in honor of the 1910 apparition of Comet Halley. Center, the heavenly porter. This cartoon by Louis M. Glackens shows a comet with the head of a sleeping-car porter and a large clothes brush for a tail, getting ready to brush off the Earth. From Puck, *Vol. 67, May 18, 1910, front cover (photographs courtesy of Ruth Freitag).*

1910 French postcard (from the collection of D. K. Yeomans).

in the amount and energy of the subatomic particles that the Sun heaves into space—the solar wind. By studying comet tails the German scientist Ludwig Biermann in 1951 demonstrated that the solar wind had to exist.

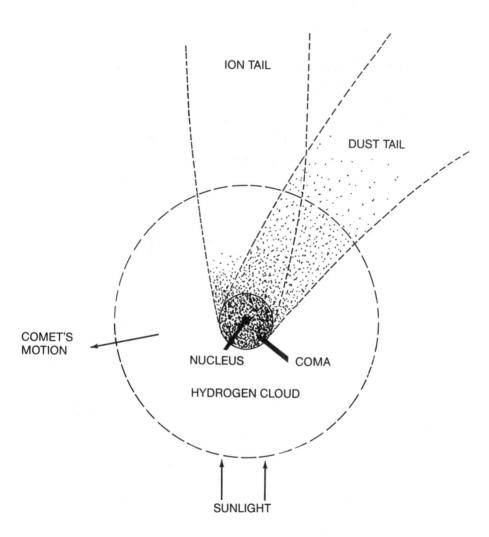

Anatomy of a comet (not to scale).

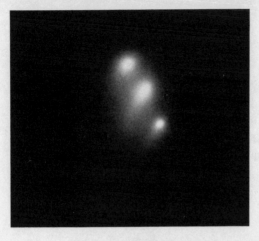

*Comet West (1976 VI)
split into four fragments
as it rounded perihelion
in February 1976. On May 26, 1976,
when this photograph was taken,
three of the four pieces were apparent
(photograph courtesy of H. L. Giclas,
Lowell Observatory).*

THE BREAKUP OF COMETS

Cometary astronomer Zdenek Sekanina has made a study of twenty-one comets that have been observed to split into fragments. Sekanina finds that of the nineteen split comets having orbits that could be classified, five were fresh comets direct from the Oort Cloud, three were fairly new, seven were older comets that had made several passes around our Sun, and four were short-period comets that had made hundreds, perhaps thousands, of passes around the Sun. As with the observed cometary brightness flares, little connection can be made between the splitting of a comet and its orbital location. A comet is just as apt to split at large distances from the Sun as at small solar distances, and it is just as likely to split before perihelion as after. The Great September Comet (1882 II) and Comet Ikeya–Seki (1965 VIII) made very close solar passages and may have been actually pulled apart by the unequal gravitational attraction of the Sun upon the near and far side of the comet's nucleus. Our Moon raises tides on Earth by a similar unequal attraction of the waters on the Earth. A third comet, short-period Comet Brooks 2, may have suffered a similar dismemberment when it passed very close to Jupiter in July 1886.

Like cometary brightness flares, the splitting of comets is not well understood. Comet experts have suggested that fragile comets, like a pinwheel, can spin faster and faster because of the rocket effect of the outgassing nucleus. When the comet's rotation rate is fast enough, the nucleus will tear itself apart. Some astronomers have suggested that cometary splitting is due to explosive chemical reactions. Yet, as British astronomer David Hughes has pointed out, one would then expect comets to preferentially split near the Sun where the Sun's heat can speed the reactions. However, comets do not tend to split near the Sun. As with the flaring of comets, we just do not know why comets split.

We have followed Comet Halley through one full 7.6-billion-mile (12.3-billion-kilometer) circuit as it traversed the solar system from outside the orbit of Neptune to inside the orbit of Venus. With each pass by the Sun, Halley's Comet loses the top meter or so of its primitive surface. Comet Halley and all the comets spend most of their time far from the Sun in a cosmic deep-freeze. This factor makes comets so valuable to science.

Of all the objects orbiting the Sun, a comet is the closest approximation we have to the original matter from which everything in our solar system was formed.

Donati's comet over Paris in 1858. Reports from that time call it one of the most beautiful comets ever seen. The dust tail curves away from the straight ion tails (from the collection of D. K. Yeomans).

Chapter Five

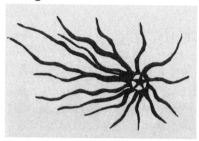

VISITING A
CELESTIAL MUSEUM

> *...now we know the sharply veering ways of comets, once*
> *A source of dread, no longer do we quail*
> *Beneath appearances of bearded stars.*
> —EDMOND HALLEY
> "Ode to Newton" (translated from Latin
> by Leon Richardson)

THE GREAT COMET CHASE

A comet is a traveling museum bringing us primordial material from the birth of the solar system. Comets glide within our reach carrying crucial chemical information about the origin of the Sun, our planet, and ourselves.

Because of these secrets comets carry, mankind is no longer content to watch them from afar. In 1986, for the first time on its journey through the centuries, Halley's Comet will have visitors. A multinational fleet of spacecraft en route to the comet will search out information unobtainable from Earth.

Japan's Planet A will fly by Halley's Comet at a distance of approximately 120,000 miles (200,000 kilometers). Its principal mission will be to study the effect of the solar wind on the comet, especially the hydrogen cloud that envelops its nucleus.

Another Japanese spacecraft, named Sakigate ("Pioneer") after launch, will study the solar wind "upstream" of Comet Halley (between the Sun and comet) by flying 4.3 million miles (7 million kilometers) from the comet at the same time as the other

THE RECENTLY DISCOVERED INGREDIENTS OF A PRIMORDIAL SOUP

One of the primary reasons for the great interest in studying comets from space is to gather clues concerning the origin of our solar system. Comets are thought to be the most primitive remnants of the chemical soup from which our solar system formed some 4.6 billion years ago. Because they have spent most of their lifetimes far from the Sun in the deep-freeze of space, they have not suffered the erosion processes or modification by solar radiation that have reshaped the interiors and surfaces of the larger asteroids and planets in the inner solar system. Because they are small, they have been spared the effects of strong self-gravity. Thus, of all the bodies remaining in the solar system since its origin, the comets have changed least. If we wish to study the chemical mixture from which our solar system originally formed, we must determine the chemicals that make up the nuclei of primitive comets.

Increasing evidence suggests that the principal ingredients in comets are water and dust. Recent ultraviolet spectral observations of comets have shown enormous hydrogen clouds, more than one million kilometers in extent, surrounding active comets. These ultraviolet observations were first made of Comet Tago–Sato–Kosaka (1969 IX) and Comet Bennett (1970 II) by two Earth-orbiting observatories, OAO–2 and OGO–5; ultraviolet observations are only possible above the Earth's atmosphere. At much longer wavelengths, radio spectral observations of many comets have established the strong presence of the hydroxyl radical (OH). If hydrogen (H) and the hydroxyl radical are observed to be abundant in comets, it makes sense to suppose that both of these chemical constituents were formed when sunlight dissociated water molecules ($H_2O \rightarrow H + OH$). The evidence, however, was not conclusive until just a few years ago when the technology to make radio and ultraviolet spectral observations became available. Most of the information on cometary ingredients has been limited to measurements made from Earth of the spectral wavelength region visible to our eyes. In the visible region of the spectrum, the strongest spectral lines are usually due to radicals such as CN, CH, C_2, and C_3. These "daughter" radicals are thought to be due to the dissociation products of still uncertain "parent" molecules such as methane (CH_4) and ammonia (NH_3). These molecules, as inferred from the visible spectrum, are now thought to make up only a small

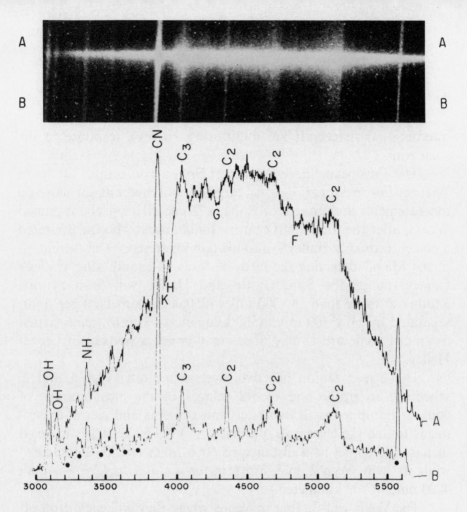

Spectra of short-period Comet Stephan–Oterma taken on November 4, 1980. Spectra A and B represent the comet's nucleus and coma region, respectively. Those spectral features with black dots below them are terrestrial, not cometary, in origin (photograph courtesy of Stephen Larson, University of Arizona).

percentage of the comet's ice content; the rest of the ice is thought to be water.

The presence of dust has been obvious for a number of years because of the telltale continuous spectrum seen in many comets. The continuous spectrum is due to the reflection of sunlight from the cometary dust particles. Thus, until recently, we have been able to see only the dust and the small percentage of molecules that have spectra in the visible region.

Astronomer Paul Feldman has noted that for millennia, what we have been observing as the spectacular phenomena of comets have not been the products of the major cometary ingredient, but only the "dirt."

international spacecraft are monitoring Halley's response to the solar wind.[1]

The European Space Agency (ESA), a consortium of eleven Western European nations, has targeted its spacecraft for close-up measurements and images of Halley's Comet. This probe is named Giotto, after the fourteenth century Italian artist who incorporated a comet, probably Halley's, into his famous fresco, "The Adoration of the Magi," depicting the birth of Jesus. A month after Halley's Comet rounds the Sun, Giotto and Halley will flash by one another at more than 150,000 miles (250,000 kilometers) per hour, separated by only 300 miles (500 kilometers). Giotto, reincarnated as an electronic artist, may once again create a portrait of Comet Halley.

The Soviet Union has two large spacecraft, Vega 1 and 2, scheduled to probe and record images of the most famous of comets. Equipped with two television cameras and eleven experiments to analyze chemicals, particles, and fields, Vega 1 is targeted to pass the comet at a distance of 6000 miles (10,000 kilometers) while its twin, Vega 2, may move in for a closer look at less than 4000 miles (6000 kilometers).

The Vegas will fly first to Venus where they will each drop off a lander to analyze the soil and an instrumented balloon to probe the Venusian atmosphere. In exchange, the Vegas will receive a gravitational assist from Venus to boost them on to Comet Halley. The Soviet spacecraft are named Vega because of their dual mission to *VE*nus (Venera) and *HA*lley's Comet. The Russians have no "H" sound in their alphabet and use a "G" sound to begin Halley's name instead. The Vega designation also provides a nice wordplay with the bright star Vega.

While four spacecraft (five if you count Sakigate) fly close by Halley's Comet, professional and amateur astronomers on Earth will follow the comet with thousands of telescopes. And Comet Halley will also be observed by astronomers in space as they orbit the Earth on a special flight of Space Shuttle. Astro-1, a special pallet of instruments for observing Comet Halley, will accompany the astronauts.

*Japanese Planet A spacecraft
(photograph courtesy of K. Hirao,
ISAS, Japan)*

*Soviet Vega spacecraft
(photograph courtesy of Vega project,
Moscow)*

PATHFINDERS IN SPACE

The European Space Agency's Giotto spacecraft to Comet Halley has a problem. To allow the on-board science instruments to observe the comet's nucleus and neighboring atmosphere to their best advantage during its flyby on March 14, 1986, the spacecraft has to pass on the sunward side and within 500 kilometers (300 miles) of the nucleus. But ground-based measurements will allow an accurate prediction of the comet's position only to 500–700 kilometers, and Giotto cannot navigate itself. If the spacecraft is aimed to ensure its passage within 500 kilometers of the comet's nucleus, it could wind up on the nighttime side of the nucleus and the cameras wouldn't see a thing. If the spacecraft is aimed to pass 1000–1500 kilometers on the sunward side of the nucleus, the on-board cameras would be ensured of a well-lit nucleus for taking good pictures, but some of the gas and dust experiments would suffer because there may not be enough gas and dust to sample at these greater distances from the nucleus.

An unprecedented international effort is now underway to solve Giotto's problem. The two Soviet Vega spacecraft will arrive at Comet Halley on March 6 and 9 of 1986, several days before the Giotto spacecraft arrives. The Vega spacecraft will not attempt to fly as close to the comet as Giotto, so a wrong-side passage is not as much of a worry. However, if the Vega spacecraft can provide information on the comet's actual position in space, this information can be used to improve the aiming accuracy of the approaching Giotto spacecraft.

For this "pathfinder" concept to work, there must be close cooperation between the separate Halley spacecraft projects in the Soviet Union and Europe and with America's NASA Deep Space Tracking Network and the International Halley Watch's Astrometry Network. The first step is for the Astrometry Network to provide the Europeans and the Soviets with Halley position data and comparison orbits that will allow them to compute the most accurate Halley orbit possible from ground-based data. The NASA Deep Space Tracking Network will track the Soviet spacecraft with respect to quasars that have well-known positions in the sky. Quasars are galaxies in formation and are extremely distant sources of radio waves. Very accurate tracking of the Soviet spacecraft is possible because of an interferometric technique that differentiates similar wavelength signals from the spacecraft and quasars that appear to be nearby in the sky. The Soviets will be tracking their own spacecraft as well, but not to the accuracy required for the pathfinder concept to work. NASA's Jet Propulsion Laboratory in Pasadena, California, will use the Vega and quasar tracking data to compute accurate positions of the Vega spacecraft at Halley encounter and then send this information to both the Europeans and the Soviets. The Europeans will take the Soviet information on the comet's position with respect to the Vega spacecraft during their flybys, combine that with the accurately determined positions of the Vega spacecraft with respect to the Earth (supplied by the Jet Propulsion Laboratory), improve the predictions of Comet Halley's position in the sky, and then perform one last maneuver of their Giotto spacecraft to enable it to fly closely past the comet's nucleus on the sunward side.

A great deal of international cooperation taking place within a very short time period will be required to make pathfinder work as planned. However, in November 1984 all the necessary international technical agencies formally agreed to cooperate in the hope of making the pathfinder concept a success.

Following page, an artist's conception of the European spacecraft Giotto as it encounters Comet Halley (photograph courtesy of the European Space Agency).

Probing Halley's Comet close up is a once-in-a-lifetime opportunity, but a proposed American spacecraft to Halley's Comet was canceled for lack of funds.[2]

Yet scientists and engineers do not easily give up their desire to know. They discovered how to send a spacecraft to a comet without building one for the purpose.

A COMET FOR COMPARISON

In orbit between the Sun and Earth was the International Sun–Earth Explorer 3, hard at work since 1978 analyzing the fast-moving charged particles of the solar wind.[3] Even before it was launched, Robert Farquhar, chief mission architect, recognized that ISEE-3 could be useful for comet studies if it could be rerouted. He worked out a remarkable series of maneuvers—orbital acrobatics—that would send this craft on a new mission for which it was never intended but for which its equipment was well suited. In June 1982, Explorer was nudged out of its Sun–Earth monitoring orbit and sent toward our Moon. Repeated lunar swingbys over a period of one year radically adjusted the spaceprobe's orbit so that on its fifth encounter in December 1983, it passed the Moon from behind and at an altitude of only 74 miles (120 kilometers).[4] The Moon's gravity and orbital velocity gave the spacecraft added speed and a new direction toward Comet Giacobini–Zinner, a comet of moderate activity that orbits the Sun every 6½ years.

As the Moon's gravity deflected the International Sun–Earth Explorer 3 onto a new course, the spacecraft received a new name: International Cometary Explorer—ICE—a very accurate description of the object to be explored.

On September 11, 1985, six months before the Comet Halley probes reach their target, ICE will fly through the tail of Comet Giacobini–Zinner at a distance of 6200 miles (10,000 kilometers) from the nucleus at a speed of 12.8 miles (20.7 kilometers) per second. ICE will be the first spacecraft ever to fly close to (and "through") a comet.

The entire mission will cost less than three million dollars, almost all of it for modifying radio receivers on the ground that will also be used for other space missions.

ICE carries no cameras, but six of its thirteen original instruments—the magnetic field and solar particle detectors—are capable of demonstrating how the icy nucleus of a comet reacts to the radiation and particle bombardment of the Sun. The camera-equipped spaceprobes headed for Halley's Comet must all pass on the sunward side to take pictures. Only ICE is scheduled to sample a comet's tail "downstream" from the Sun.

Information from ICE about the gas and dust environment near its comet should help Japanese, European, and Soviet space scientists know what to expect from Comet Halley. But most valuable is that Giacobini–Zinner is a different comet and therefore provides an opportunity for comparison. Planets have differences. Comets do, too.

Chapter Six

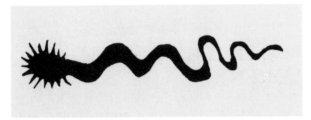

PROSPECTS FOR VIEWING HALLEY'S COMET

Of all the comets in the sky,
There's none like Comet Halley.
We see it with the naked eye,
And periodically.
—HAROLD SPENCER JONES

EXPECTATIONS

Because Halley's Comet has such scientific importance and is so dependable, preparations for its return were made years in advance to allow for spacecraft visits. Also, thousands of professional and amateur astronomers throughout the world have joined together in the largest program of international cooperation ever undertaken in astronomy, the International Halley Watch.

The International Halley Watch, with Western Hemisphere headquarters at the Jet Propulsion Laboratory in Pasadena, California, will also serve as a clearinghouse for the latest information on Halley's Comet.[1]

Halley's Comet is the most famous of its breed, but not the brightest. Comet Ikeya–Seki in 1965, Comet Bennett in 1970, and Comet West in 1976 were all brighter. But they will not return for many hundreds of years.[2] Comet Halley is our guest approximately every seventy-six years and is the brightest of all the short-period comets.

A TREAT FOR OUR GREAT-GREAT GRANDCHILDREN

For most people, the 1985–86 return of Comet Halley will undoubtedly be a disappointment if they rely on naked-eye observations from the city. For much of the time when the comet is at its brightest, it will be on the opposite side of the solar system from our Earth and hence impossible to observe. Of course, the spacecraft observations and optical aids available to ground-based observers will provide an enormous amount of scientific data on Comet Halley for astronomers and enjoyable pictures for all. For those who know where, when, and how to look for the comet, there will be satisfying personal encounters. But for most people, Comet Halley won't be very impressive this time around.

According to computations by D. K. Yeomans, after 1986, the comet's next return to perihelion will be on July 28, 2061. During the few months prior to perihelion passage in 2061, the viewing circumstances will not be very good. But after perihelion, during early August 2061, the comet should be a bright object to the naked eye. At apparent magnitude 0 to 2, it may be approximately five times brighter than in 1986. Ask your grandchildren to mark their calendars.

You may also urge your grandchildren to inform their grandchildren about Comet Halley's return to perihelion in 2134. After the comet passes perihelion on March 27, 2134, it will pass within 0.093 astronomical units (8.6 million miles; 13.9 million kilometers) of the Earth on May 7, 2134, approximately three times closer than Venus ever gets. It should achieve an apparent magnitude of nearly –2, brighter than Sirius, the brightest of the nighttime stars. With a tail stretched across the sky, the comet should put on its best show since its closest recorded approach to the Earth on April 10, A.D. 837, when the Earth–comet distance was only 0.033 astronomical units.

Comet Halley as seen on June 6, 1910 (photograph by E. E. Barnard courtesy of Yerkes Observatory)

Below, Halley's Comet in 1910 as observed from April 26 to June 11. These photos reveal the changing apparent length of the tail (photograph courtesy of Mt. Wilson and Las Campanas Observatories of the Carnegie Institution of Washington). Opposite, Comet Ikeya-Seki (1965 VIII) was one of the brightest comets of the twentieth century (photograph by C. F. Capen, Planetary Library and Historic Chart Archive).

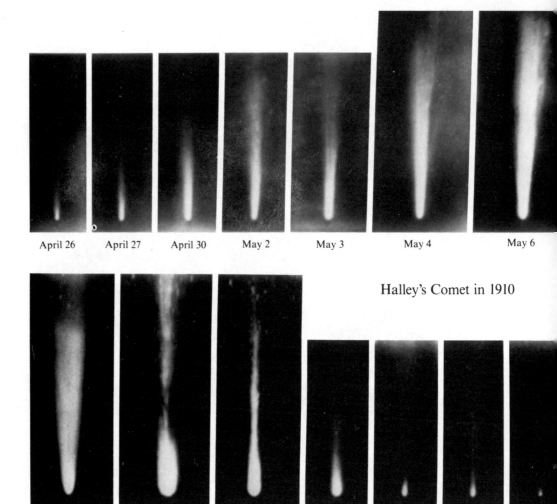

| April 26 | April 27 | April 30 | May 2 | May 3 | May 4 | May 6 |

| May 15 | May 23 | May 28 | June 3 | June 6 | June 9 | June 11 |

Halley's Comet in 1910

In late March and early April 1986, as Halley's Comet is outbound from the Sun and approaching Earth, some astronomers think it may reach magnitude +2—approximately the brightness of Polaris, the North Star.[3] On a clear, dark night away from city lights, Halley's Comet should look like an average starlike dot

THE INTERNATIONAL HALLEY WATCH

by **Ray L. Newburn, Jr.**
Leader
International Halley Watch

In 1910, when Comet Halley last visited the inner planetary neighborhood, astronomy was just entering the modern era. The largest telescope in the world was the 60-inch reflector on Mt. Wilson, finished just two years earlier. Photography had become quite routine, but astronomers still made some visual measurements.

After Halley's Comet was spotted inbound to the Sun on September 11, 1909, a group of astronomers formed "The Comet Committee" to organize studies of the comet worldwide. The committee recommended "a photographic campaign as long and as nearly continuous as possible" to record changes in the comet's head and tail. They also proposed spectroscopic, photometric, and polarimetric observations. The committee specifically asked for a copy of each photograph taken so that members personally could compare the collected results.

Very little was actually accomplished by the Comet Committee. They finally published a list of photographs, noting that most were taken at just a few observatories that planned their own publications. The two major observatory monographs on Comet Halley did not appear until 1931 and 1934. Much of the data acquired was never published anywhere. (Attempts are being made to publish some of it in 1985.) The problems of 1910 included a late beginning, poor communication, and inadequate resources.

The International Halley Watch (IHW) was organized in 1980 in recognition of the potential value of a comet study coordination center and with full awareness of the problems of 1910. The differences between 1910 and 1986 are numerous, and not just technological. The IHW recognizes that each astronomer has the right (and duty) to publish his own findings. It asks only that copies of the organized data be transmitted to the IHW for inclusion in a Halley Archive to be published in 1989. That archive, probably in the form of digital video disks as well as books, will be made available to all researchers who contribute to it. Meanwhile, the IHW regularly supplies to participants accurate ephemerides, standards, observing suggestions, and other forms of support. The scientist who participates has everything to gain and nothing to lose. By the end of 1984, 886 professional astronomers and 314 amateur astronomers in 47 countries were working with the IHW. Besides the professional support it provides, participation in the IHW offers the sheer fun of being part of a worldwide activity of great scientific value.

Modern science has armed researchers with detectors that are hundreds of times more sensitive than the photographic plate—detectors that "see" in all regions of the spectrum, not just visual light. During this apparition, Comet Halley will be studied by every possible technique, and often at the same time, on prearranged Halley Watch days.

In support of the European, Soviet, and Japanese spaceprobes scheduled to intercept Comet Halley, the IHW is supplying critical data on the comet's position so that the spacecraft can be guided to the proper location in space for their encounters. The spacecraft data tapes ultimately will be included in the Halley Archive.

The spacecraft will all race by Halley's Comet during the short period of March 6–14, 1986. It will be the responsibility of ground-based astronomers, observing Halley's Comet since it was recovered in 1982, to place the spaceprobe results into the context of the entire period of Halley's activity.

The International Halley Watch, then, is the ground-based part of an unprecedented worldwide effort to better understand the most famous of comets. The rareness of the opportunity has brought out the very best in science and in scientists on every continent.

surrounded by a fuzzy blur and with a faint tail stretching away from the Sun's direction. No wonder the Greeks called these objects "hairy stars" *(kometes)*.

How well we see Halley's Comet depends on where the Earth is as the comet swings around the Sun. It also depends on the following:

- *How bright the comet actually becomes.* Comet brightness is hard to predict because it is impossible to know the changing ratio of the ice–dust mixture at the surface of a comet and hence how much material will evaporate. It is impossible to know when chunks of the comet's surface will break loose and briefly enhance the comet's brightness. It is impossible to know when a small meteoroid will strike the comet's solid nucleus and create an impact explosion that throws gas and dust off the comet and causes it to brighten for a time. For example, in the fall of 1984 Comet Halley was already brighter than expected because vaporization had given the comet a coma even though the comet was still beyond Jupiter. This early brightening of Halley's Comet does not mean that the comet will continue to be brighter than expected; it only means that comet brightness is tricky to predict.

- *How bright your city lights are.* The dimmer the lights or the farther away from town you are, the better. (The comet may not be visible to the unaided eye from cities.)

- *How severely your skies are polluted.* Again, the farther from town you are, the better.

- *How bright the Moon is.* The less full the Moon is, the better.

- *How far south you live.* The farther south you live, the better. For observers in Canada and the northern United States, the comet will be near the horizon through its period of maximum brightness from late March through mid-April 1986.

Comet West (1976 VI) was the brightest comet so far this century (photograph by Dennis diCicco, Sky & Telescope Magazine*).*

Because for us on Earth Halley's Comet will not be as bright in 1985–86 as it was in 1910 and on some of its earlier visits to our neighborhood, a telescope or binoculars will be especially useful. Whether you look for the comet with your eyes alone or with optical aid, allow ten to twenty minutes for your eyes to become adapted to the dark.

Remember also that comets are not meteors. They do not flash across the sky in a second or two. Comets are moving rapidly, but they are millions of miles away and therefore change their positions among the stars rather slowly.

November 1985

A small telescope or even binoculars should bring Halley's Comet within view by mid-November 1985 when it will be in Taurus,

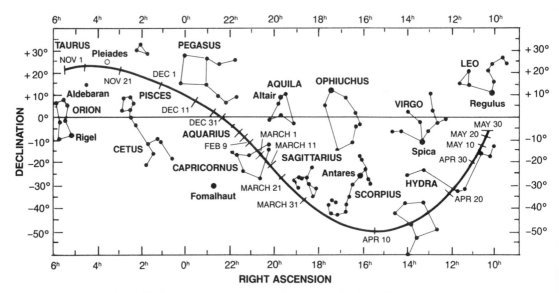

Path of Halley's Comet among the constellations from November 1985 through May 1986 (IHW chart from The Comet Halley Handbook, *NASA).*

rising in the east in the early evening. Consult the International Halley Watch star chart (p. 112) for the comet's position among the stars. On the night of November 16–17, Halley's Comet will be just south of the Pleiades star cluster. The comet's head should be small and fuzzy, and its tail should be short and faint. Telescopic viewing of comets is best with low-power (wide-field) eyepieces.

On November 27, Comet Halley, inbound toward the Sun, will pass the Earth at a distance of 57.6 million miles (92.8 million kilometers), but bright moonlight will hinder viewing both early and late in the month.

December 1985

In early December, when Halley's Comet is high in the southern sky at nightfall, it may become bright enough to be seen in the evening sky with the unaided eye if you are away from city lights and know exactly where to look. Find the comet in Pisces with binoculars or a telescope first to fix its position and enjoy its fuzzy coma and growing tail. Halley's Comet will remain on the border of naked-eye visibility throughout the month as it glides closer to the Sun but farther from Earth.

January 1986

Through a telescope especially, the comet should look like a little star surrounded by a haze, with a small tail extending from it. The comet, visible in the western sky in Aquarius after dark, will still be very faint, detectable by the eye alone only under the best (darkest) observing conditions. By the end of the month, the comet goes out of sight beyond the Sun.

February 1986

Halley's Comet is closest to the Sun (54.5 million miles; 87.8 million kilometers) on February 9, but we will not be able to see it because Halley's Comet will be on one side of the Sun, and we on Earth will be on the opposite side.[4] Comet Halley will not be directly behind the Sun, but it will be too close to the Sun's position to appear in the night skies.

Because of Earth's unfavorable position when the comet is closest to the Sun and hence near intrinsic maximum brightness, our 1986 encounter with Halley's Comet will not be visually overwhelming.

Beginning around February 20, try watching for the comet's tail to rise over the eastern horizon approximately 1½ hours before sunrise. Before the comet's head can rise, morning twilight will end

HOW COMETS GET THEIR NAMES

Upon discovery, a comet is given a preliminary designation based upon its order of discovery in a particular year. Thus, at its last return to the Sun, Comet Halley was first recognized, after a three-year search, on September 11, 1909, by the German astronomer Max Wolf at the Heidelberg Observatory. It was the third comet discovered in 1909, so the preliminary designation was 1909c. After enough observations were available to establish its orbit, it was determined that Comet Halley did not reach its closest approach to the Sun (perihelion) until April 20, 1910. Because it was the second known comet to reach perihelion in 1910, it was given its final numerical designation of 1910 II. A designation like 1910 II is fine for catalogs and technical papers, but almost everyone refers to this comet as Halley's Comet. In so doing they are honoring Edmond Halley who first successfully predicted the return of this comet.

In the early nineteenth century, the German astronomer Johann Encke recognized that Comets 1819 I, 1805, 1795, and 1786 I were the same object returning to the Sun at 3.3-year intervals. This comet was named Comet Encke thereafter by everyone except Encke himself who modestly referred to it as Pons' Comet, after the French astronomer Jean Louis Pons who discovered it at Marseilles in 1805. In fact, comets are usually named after the person or persons who first discover them. Comets Halley and Encke are exceptions as are the two comets discovered by the Chinese in 1965 at the Tsuchinshan Observatory at Nanking. These latter comets are called Tsuchinshan 1 and Tsuchinshan 2 because their Chinese discoverers were too modest to have their names attached to them. Outside of China, however, the honor of having a comet named after oneself is very eagerly sought.

the viewing. The comet is now outbound, proceeding tail first away from the Sun. By perhaps February 23, the comet's head may be visible before the sunrise twilight becomes too bright. The head of Halley's Comet may now be nearly as bright as the North Star, Polaris. The comet is once again moving closer to Earth, so night

The instant immortality that comes from discovering a comet is a heady inducement to many amateur astronomers who spend hundreds of hours trying to be the first to identify a new comet. If several observers discover a new comet nearly simultaneously, up to three names can be associated with it. For example, the Japanese amateur astronomer Minoru Honda and the Czech astronomers Antonin Mrkos and Ludmilla Pajdusakova all nearly simultaneously discovered a comet in early December 1948. It is now referred to as Comet Honda–Mrkos–Pajdusakova.

Unlike comet hunters, the discoverers of asteroids are given the honor of naming their discovery virtually anything they wish. A problem arises because so many asteroids are discovered each year, but only a handful of astronomers are making the discoveries. Hence, after naming asteroids for their scientific colleagues, past and present, the relatively few people making asteroid discoveries must turn to less and less appropriate designations just to give each one a different name. Thus, among the numbered asteroids that have known orbits, we find ones named after cities (2200 Pasadena), buildings (2288 Karolinum), restaurants (2038 Bistro), and even airlines (2138 Swissair). For those with prurient interests, we have 433 Eros, 2063 Bacchus, and 2174 Asmodeus, which are named after the gods of love, wine, and lust, respectively.

The wide variety of names for the more than 2000 cataloged asteroids opens up exciting possibilities for sheer nonsense. If one has absolutely nothing better to do, one can form phrases by putting together various asteroid names. For example, if you are feeling nasty, you could put together asteroids 304, 1485, and 426 (Olga Isa Hippo), or if your mood is more whimsical, the combination of asteroids 904, 673, 449, 848, and 1136 will give you Rockefellia Edda Hamburga Inna Mercedes.

by night the comet's tail will seem longer, perhaps extending five degrees, the equivalent of ten diameters of the full Moon. Use binoculars or a telescope to search for detail in the coma and tail.

March 1986

Comets are often brightest and their tails longest soon after they have basked in the warmth near the Sun, so the best time for viewing Halley's Comet will come soon after it begins its outward journey from the Sun and especially so because it will then be passing closest to Earth.

By mid-month, there will be no moonlight to interfere with viewing Halley's Comet in the southeast before morning twilight appears. The comet's head will still be a little fainter than Polaris, but the comet's tail will now extend ten to fifteen degrees, the equivalent of approximately one-eighth of the distance from the horizon to the top of the sky (zenith). By March 25, the tail may be nearly twenty degrees long. Possibly you will see stars shining undimmed through the comet's tail.

April 1986

Halley's Comet should reach maximum brightness as seen from Earth during the period April 5–10. But the comet's outbound orbit carries it southward. If you are an observer living in Canada or the northern half of the United States, Comet Halley will now be below or too close to the southern horizon for viewing. The best views will be from Hawaii, southern Florida, southern Texas, and farther south to well below the equator in early April. As you go south, the comet is higher in the sky.

Near the end of the first week in April, moonlight will no longer be in the way, and the comet's coma may appear half the diameter of the Moon. You should notice that the comet has two tails: a faintly bluish straight tail made of ionized gases and a curved yellowish-white tail made of dust. The dust tail could now stretch more than one-fourth the height of the sky.

On April 11, Halley's Comet passes its closest to the Earth this trip: a distance of 39.0 million miles (62.8 million kilometers). If you are in the southern United States or farther south as the comet sails by us, you will see it from changing angles. Watch over a period of days as the comet's tail, always extending away from the Sun, pivots from northwest to north to northeast. But the comet is now getting fainter.

After the middle of April, as Halley's Comet recedes from Earth, it switches from a morning to an evening object. We can see it again from the northern United States and Canada in the southeastern sky after nightfall.

May 1986

The comet will now be only as bright as the faintest star in the Big Dipper. Your binoculars or telescope will be increasingly essential to follow this once-in-a-lifetime visitor. By the end of May, the comet will no longer be visible to your eyes alone.

June 1986

By consulting the International Halley Watch star chart (p. 112) for the position of Halley's Comet among the stars, you should be able to follow the comet with your telescope and binoculars all month, and perhaps beyond.

THE COMET HALL OF FAME

Physical Features

Longest tail: 2.15 astronomical units—Comet 1843 I.

Longest "antitail": fourteen degrees—Comet Arend–Roland (1957 III), April 24, 1957.

Most tails: six or seven—Daytime Comet of 1744 (De Cheseaux's Comet).

Brightest comet (as seen from Earth): Daytime Comet of 1744 (De Cheseaux's Comet)—visible only twelve degrees from the Sun on February 25, 1744; magnitude approximately –5.

Comet with the greatest intrinsic brightness: Comet of 1729— although it had a perihelion distance of 4.1 astronomical units, it was a naked-eye object on July 31, 1729, when the Earth–comet and Sun–comet distances were 3.1 and 4.1 astronomical units, respectively.

Brightest short-period comet: Comet Halley.

Comet that passed closest to Earth without colliding: Comet Lexell (July 1, 1770)—1.4 million miles (2.3 million kilometers); its coma was two degrees and forty minutes, five times the apparent diameter of the full Moon.

Shortest orbital period: Comet Encke—3.3 years.

Most apparitions: Comet Encke—it made its fifty-third observed return to the Sun in 1984.

Smallest perihelion distance: Great Southern Comet (1887 I) discovered by Juan (John) Thome—0.0048 astronomical units from the center of the Sun (only approximately 14,000 miles or 23,000 kilometers above the surface of the Sun). Several comets have been known to hit the Sun.

Comet that split into the most pieces: Great September Comet (1882 II)—five pieces.

Comets having tails through which the Earth has passed: Comet Tebbutt (1861 II)—June 29–30, 1861; and Comet Halley (1910 II)—May 19, 1910.

Comet Kohoutek (photograph courtesy of Palomar Observatory)

Most distant comets observed: outbound: Comet Stearns (1927 IV)—11.53 astronomical units; inbound: Comet Halley (October 16, 1982)—11.04 astronomical units.

Faintest comet observed: Comet Halley (October 16, 1982)— magnitude 24.2.

Only short-period comets with retrograde orbits: Comets Halley, Tempel–Tuttle, Pons–Gambart, and Swift–Tuttle (out of more than 100 short-period comets known); by contrast, approximately fifty percent of long-period comets have retrograde orbits.

Comets and Mankind

Most comets discovered (or codiscovered) by a single person: thirty-six or thirty-seven—Jean Louis Pons (1761–1831); he discovered his first comet in 1801.

Comet Kohoutek (photograph courtesy of Palomar Observatory)

THE COMET HALL OF FAME—*cont.*

Most comets discovered in a year: five—Jean Louis Pons (in eight months: February–September 1808); Jean Louis Pons (in just over twelve months: August 1826–August 2, 1827); Francesco De Vico (1846); and William H. Brooks (1885–86).

First proof that comets are extraterrestrial: Tycho Brahe, using parallax measurements on the Comet of 1577, determined that this comet lay beyond the Moon.

First suggestion that comets travel on parabolic trajectories with the Sun at one focus: Georg Samuel Dorffel, a German astronomer, in 1681.

First successful prediction of a comet's return: Edmond Halley in 1705 identified the comets seen in 1531, 1607, and 1682 as the same object and calculated that it would return again in 1758.

First recovery of a comet based upon a previous prediction: Johann Georg Palitzsch, a German amateur astronomer, recovered Halley's Comet on Christmas night 1758.

First comet to be observed telescopically: Comet of 1618 by Johann Baptist Cysat of Switzerland and John Bainbridge of England.

First comet discovered by telescope: Great Comet of 1680, by Gottfried Kirch, Coburg, Germany, on November 14, 1680.

First photograph of a comet: Comet Donati (1858 VI) by a portrait artist named Usherwood on September 27, 1858.

Earliest extant photograph of a comet: Comet Tebbutt (1881 III) by P. J. C. Janssen, June 30, 1881.

First photograph of Halley's Comet: Knox Shaw at the Helwan Observatory in Egypt on August 24, 1909; the image was not recognized as Halley's Comet until later.

First comet discovered by photography: Comet 1892 V, by Edward Emerson Barnard, October 12, 1892. (A photograph of the solar corona taken during an eclipse of the Sun on May 17, 1882, showed a comet, but it could not be found subsequently.)

First comet examined by spectroscope: Comet Donati-Toussaint (1864 III).

First comet to have its spectrum recorded by photography: Comet Tebbutt (1881 III).

Most maligned comet: Comet Kohoutek (1973 XII)—it didn't become as bright as originally expected, but its discovery at a great distance allowed astronomers time to plan extremely valuable observations.

First woman to discover a comet: Caroline Herschel (sister of William Herschel, discoverer of Uranus)—August 1, 1786 (Comet 1786 II).

Spectra of short-period Comet Encke taken on November 4, 1980 (photograph courtesy of Stephen Larson, University of Arizona)

First astronomer to give his life trying to find a comet: Ernst Friedrich Wilhelm Klinkerfues, who had previously discovered six comets, fell from the observing platform at the Göttingen Observatory in Germany on January 28, 1884.

Award for successful predictions under difficult circumstances: Jean-Baptiste-Gaspard de Saron (1730–1794), last Premier Président of the Parlement de Paris, successfully predicted the postperihelion motion of Comet 1793 I. This prediction allowed Messier to recover the comet after it came out of the Sun's glare. Saron's complex computations were made all the more difficult because he was in jail awaiting the guillotine during the French Revolution.

Most horrifying description of a comet that was probably not a comet: Ambroise Paré (pioneer French surgeon) on the "Comet" of 1528:

This comet was so horrible, so frightful, and it produced such great terror in the vulgar, that some died of fear and others fell sick. It appeared to be of excessive length, and was of the color of blood. At the summit of it was seen the figure of a bent arm, holding in its hand a great sword, as if about to strike. At the end of the point there were three stars. On both sides of the rays of this comet were seen a great number of axes, knives, blood-colored swords, among which were a great number of hideous human faces, with beards and bristling hair.

There are no other reports of this object as a comet. Perhaps it was a display of the aurora (northern lights). In 1528, Paré was eleven years old; his recollection was written many years later.

Soon To Be Added?

First comet passage by a spacecraft: Comet Giacobini–Zinner by International Cometary Explorer (United States), September 1985.

First comet nucleus to be imaged: Comet Halley by Giotto spacecraft (European Space Agency), March 1986.

HALLEY'S COMET POSITION CHARTS

*The dates marked on the five Halley's Comet position charts
that follow show the comet's position at the beginning of
morning astronomical twilight or at the end of evening
astronomical twilight (usually about 1½ hours before sunrise
or 1½ hours after sunset). The numbers in parentheses
indicate the comet's expected magnitude on those dates: the
lower the number, the brighter the comet. (Polaris, the North
Star, is magnitude +2.0.) Viewing with binoculars and ideal
observing conditions are assumed (IHW charts from* The
Comet Halley Handbook, *NASA).*

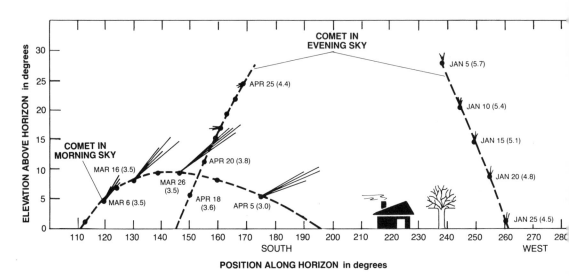

Halley's Comet at 40° north latitude.

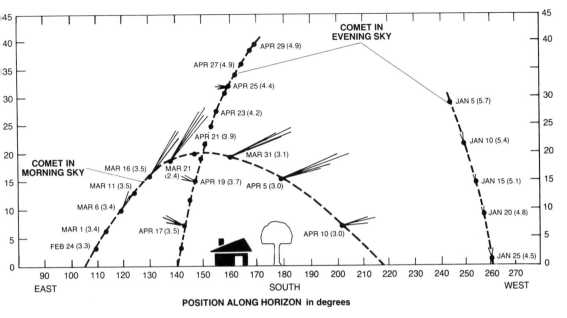

Halley's Comet at 30° north latitude.

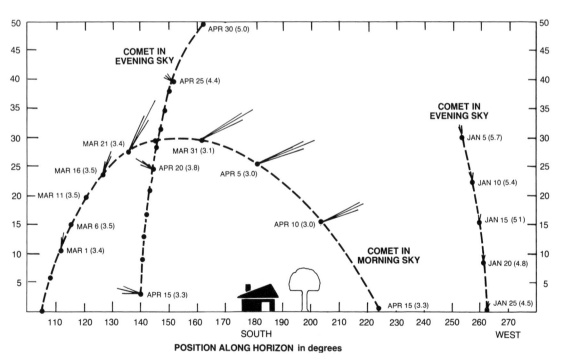

Halley's Comet at 20° north latitude.

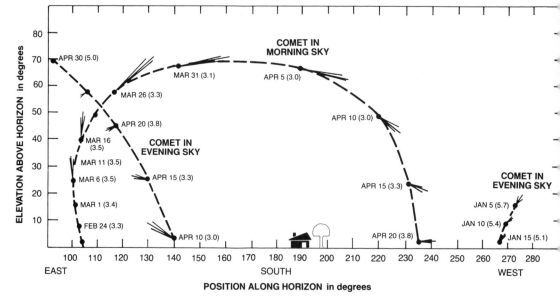

Halley's Comet at 20° south latitude.

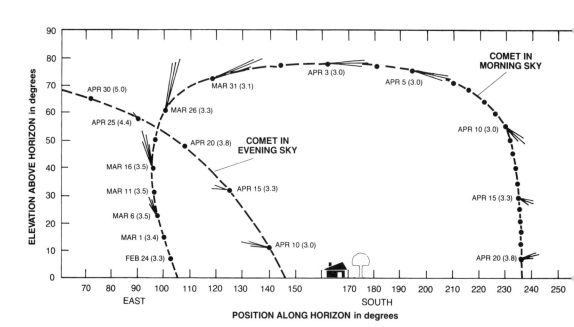

Halley's Comet at 30° south latitude.

Chapter Seven

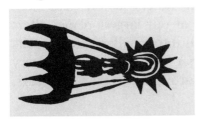

DESCENT FROM AN ORBITING AVALANCHE

> *...the great wild satellite has traversed its course and is speeding on its homeward journey toward the sun.*
> —LOREN EISELEY

WHENCE COMETH THE COMET?

As Halley's Comet departs from the Sun, it goes tail first, the sunlight and solar particles still driving material from the coma and tail outward.

A comet has far too little gravity ever to reclaim this gas and dust. With each visit to the Sun, the nucleus of a comet diminishes.

Halley's Comet cannot survive more than one thousand or so passages by the Sun, so it cannot have existed for more than approximately 75,000 years in its present orbit. Nor could it have formed along its present path just a few thousand years ago. The gravity of the planets long ago swept up the material along Halley's orbit from which a comet could form. Thus, Halley's Comet demands of us an explanation for its existence.

Comets that frequently pass close to the Sun cannot last long. But our solar system has existed for 4.6 billion years. And we continue to see comets with short orbital periods. So new comets must be taking up orbital residence near the Sun to replace the ones that fade out or disintegrate. Where are they coming from?

Almost all of the bright comets we see have orbits that carry them great distances beyond our outermost planet. As these far-traveling comets reach the inner solar system, a few pass close enough to Jupiter so that the gravity of our largest planet alters their courses and speeds, and they fall into smaller, faster orbits around the Sun.[1] Instead of comets that visit the Sun once every million years, they now complete their orbits in less than 200 years. They have become short-period comets.

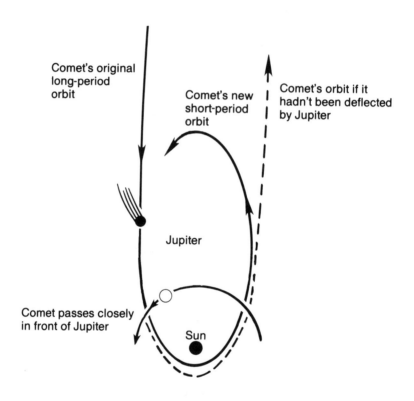

Jupiter's enormous gravity can deflect a close-passing comet from a long-period orbit to a short-period orbit (although seldom in a single encounter).

The orbits of the short-period comets have been studied, and almost all indicate a major gravitational encounter with Jupiter a few centuries or millennia ago.

The scenario for Halley's Comet is not yet fully understood. Donald K. Yeomans calculates that Halley's Comet was diverted by Jupiter from a long-period orbit to a short-period orbit not less than 16,000 years ago.

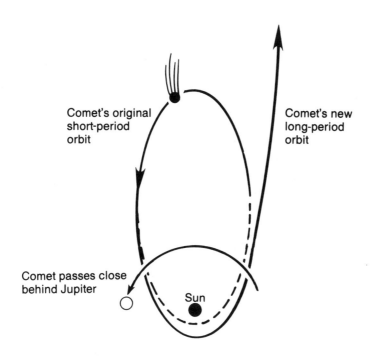

Jupiter's enormous gravity can accelerate a comet out of the solar system (although seldom in a single encounter).

SHOOTING STARS AND METEOR SHOWERS: COMETARY DEBRIS

Comets dissipate with age. Cometary gas and dust is forever lost to the nucleus whenever an icy comet nears the solar furnace. A surface layer of one meter or so is peeled away during each solar passage. Occasionally the dissipation of a comet is dramatic, as when a comet splits. More often the dissipation of a comet is gradual, and the exuded dust and debris tag along behind the parent comet. An active comet with much of its dirty ice remaining can throw out enormous quantities of dust. This debris will be relatively dense near the comet itself. The dusty debris will, after a long time, spread completely around the comet's orbit. The dust particles released by the comet can remain for 10,000 years in their own orbits near where the comet itself is, or was, traveling. This flattened tube of dust, spread around a comet's orbit, is called a meteor stream. It is referred to as a young stream if the dust particles are still bunched up near the parent comet and an old stream if the dust particles have evolved all the way around the parent comet's orbit. The same type of phenomenon occurs every Memorial Day weekend during the Indianapolis 500-mile auto race. At the beginning of the race, all the cars are bunched together. As the race progresses, the faster and slower cars separate and spread out around the entire oval raceway. Of course, a cometary dust particle traveling around its near-oval course can make a race car traveling an average of 200 miles per hour look rather slow moving. For example, the dust particles traveling in Comet Halley's orbit can reach speeds of 122,000 miles per hour at perihelion. The only pit stop made by a cometary dust particle is when it slams into the Earth's atmosphere and is consumed in fire to become a meteor or "shooting star."

Every year in early May and again in late October, some of the dust particles lost by Comet Halley bombard the Earth's atmosphere and produce a meteor shower. In early May, these shooting stars appear to originate from near the star Eta in the constellation of Aquarius. In late October, the meteors seem to stream outward from the constellation of Orion. These two modest meteor showers are called the Eta Aquarids and the Orionids after the constellations from which they appear to originate. In a good year with clear

A woodcut illustration of the remarkable Leonid meteor shower that occurred on November 12, 1799, when the Earth ran into a concentrated swarm of dusty debris from short-period Comet Tempel–Tuttle (from the collection of D. K. Yeomans).

skies and no interference from moonlight, an observer can expect to see approximately ten to twenty meteors per hour from each of these displays. Even when no meteor shower is in progress, approximately five to ten meteors per hour can be counted. These sporadic meteors are believed to be due to very old cometary dust particles that have strayed far from the orbital path of the parent comet. Very new meteor streams can occasionally cause meteor storms during which the sky rains meteors by the thousands. These magnificent displays can only occur when the Earth passes through a swarm of densely packed young meteoroids near the comet itself. The most famous example of a meteor storm is the Leonid Shower, which is due to the closely packed young dust particles associated with short-period Comet Tempel–Tuttle. According to computations by D. K. Yeomans, the Leonid storms, so impressive in 1833, 1866, and 1966, may well return again on November 17 of 1998 and 1999.

THE COMET RESERVOIR

So Halley's Comet was once a long-period comet. During its rare visits to the Sun, it encountered Jupiter and became the more frequent, faithful visitor we know today.

But long-period comets that plunge in close by the Sun cannot last long either, on an astronomical time scale. Sooner or later, they will vaporize in the Sun's heat or encounter Jupiter's gravity. Most will either be ejected from the solar system or diverted into shorter orbits where they will dissipate more quickly.

So the long-period comets we see are also doomed. An unseen supply of comets must exist beyond the most distant planet to replace those that sacrifice themselves at the altar of the Sun.

But where? Long-period comets have orbits that extend up to one thousand times farther from the Sun than Pluto.

Sunlight takes $8^{1}/_{3}$ minutes to reach the Earth. It reaches Pluto in five hours. That same light will reach the outermost comets in one year. Pluto is not the boundary of our solar system. These comets are.

But at such a distance—one light year, almost 6 trillion miles (9.5 trillion kilometers)—are comets even part of our solar system any more? The nearest star to our Sun is Alpha Centauri, which is actually a three-star system. Alpha Centauri is 4$\frac{1}{3}$ light years away. So comets one light year from our Sun would still feel our Sun's gravity more than that of the nearest stars.

Yet our Sun is always in motion through space, carrying the planets, moons, and comets with it. All 400 billion stars in our Milky Way Galaxy are in motion as well. Every once in a long while, a star or a concentrated cloud of cosmic gas and dust will pass along the fringes of our solar system, close enough for its gravity to disturb the comets.

The disturbance will cause a number of comets to accelerate out of the solar system, never to return. But the encounter will also slow some comets, causing them to fall from their age-old paths that always kept them so far from the Sun. Their new courses will carry them into the heart of the solar system. There they will eventually perish, although they will die gloriously.

In this way, stars passing close to our solar system tap the supply of unseen comets and send them toward us on Earth as replacements for all the comets that have flourished and vanished in the sunlight.

But stars pass close enough only rarely. So the supply of unseen comets must be vast to be able to provide us with the comets we see. Astronomer Paul Weissman calculates that perhaps two trillion comets reside in this comet cloud that exists at the fringe of our solar system.

If the average comet has about one-trillionth the mass of Earth, then the mass of all these comets must total about twice that of Earth.

We call this reservoir of comets the Oort Cloud, after Jan H. Oort, the Dutch astronomer who demonstrated in 1950 that this cloud of comets must exist.[2]

BACK TO THE BEGINNING

How did this orbiting avalanche of comets come into existence? The exact answer to this question is still uncertain, but its solution certainly encompasses the origin of our solar system.

The same large telescopes that searched the skies for the return of Halley's Comet show us that everywhere within our Milky Way and in other galaxies are huge clouds of gas and dust between the stars. These clouds are so dense that stars must be forming there by gravity. These gaseous nebulas are the birthplace of stars.

Nearly 5 billion years ago, in one small sector of an interstellar cloud where hundreds of stars were forming, a very average star also began to form by gravity. Near that star-to-be, stirred by turbulence within the cloud, other pockets of gas and dust began by gravity to accumulate material.

But the density was greatest at the center of this cloud fragment, and there the pressure of gravity raised the temperature so high that a nuclear reaction was sustained. Our Sun had formed. All around, other would-be stars were struggling to reach the proper size. But the particles and radiation of the newly shining Sun swept away the remaining gas and dust and blasted outward great quantities of the gas of the would-be stars themselves, treating them as if they were giant comets. The stars-to-be failed to reach the critical mass for a nuclear reaction. They could not emit light of their own, but they could reflect sunlight. They became the planets.

This theory, astronomers believe, explains how our solar system began. But as to how the comets fit in, the opinion is divided.

THE ORIGIN OF COMETS

Could it be that the comets formed far from the Sun in a comet cloud with a diameter much smaller than the Oort Cloud in which most of the comets may be found today? At the fringe of the solar nebula from which the Sun and planets formed, the material was so sparsely spread that no large planets could develop. Here, according to one hypothesis, the pressure of the light from nearby stars pressing inward and the light of our Sun pressing outward molded the cold gas and dust into the mile-wide icebergs we call comets.[3] Over millions of years, passing stars and nebulas would have warped the orbits of these comets so that many wandered outward into larger orbits, extending the Oort Cloud to a distance of approximately 150,000 astronomical units (approximately 2½ light years). Such gravitational perturbations would also have hurled these comets into orbits of random inclination and direction so that comets falling from the Oort Cloud today come from all directions.

An alternative possibility is that the comets formed much closer to the Sun where the outer planets Uranus and Neptune lie today. In that region, water is always in the form of ice. And comets are mostly water ice. The chunks of ice collided, sometimes hard enough to shatter, but sometimes gently enough so that the snowballs clung to one another, their gravitational attraction for other chunks increasing. Two giant iceballs evolved. Today we call these overgrown "comets" Uranus and Neptune. But billions of tiny icy conglomerates remained and were continually deflected into new orbits by Uranus and Neptune and then by inlying Saturn and even more powerful Jupiter.

In this early era, the giant planets used their gravity to hurl dirty snowballs at one another, at the Sun, and at the smaller planets and moons. Many billions of comets must have perished in these collisions; billions more evaporated. An equal number of comets must have been hurled outward at such speeds that they were forever banished from the solar system. But a few, perhaps two trillion, were not lost. Hurled out beyond Pluto, their lesser speed left them distributed randomly in a vast zone up to two light years away.

But whether comets formed within the Oort Cloud at the beginning of the solar system or were later flung into that region, these distant snowballs were still captives of the Sun's gravity, moving unseen and, for 4.6 billion years, unsuspected.

A COMET CALLED HALLEY

One of these icy wanderers spent all these eons in this cloud of comets until the gravity of some now unknowable passing star or nebula or our Sun's possible stellar companion sent it on its fateful

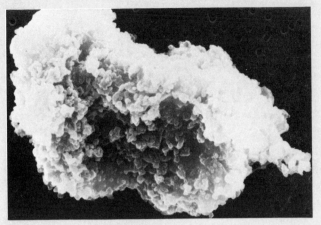

A cometary dust particle captured in March 1981 with a special particle trap carried by a high-flying U-2 aircraft. Thi. particle is only fifteen thousandths of one millimeter across and has been magnifiea 15,000 times by an electron microscope (photograph courtesy of Donald E. Brownlee, University of Washington).

THE DEATH OF COMETS

Not only do comets often appear unexpectedly, but they can also disappear unexpectedly. After it split in two in 1846, short-period Comet Biela was seen as a pair of comets in 1852 and never again thereafter (although it provided impressive meteor showers in 1872 and 1885). Short-period Comet Westphal was seen in 1852 and again in 1913, but it actually faded as it approached perihelion that year. It was not seen after perihelion, and despite searches in 1975–76 during its last expected return, it was not recovered. Short-period Comet Perrine–Mrkos provided five apparitions from 1896 to 1968, but astronomers searched in vain for it in 1975.

It is not at all clear whether comets simply disintegrate as they lose their ice and dust, or whether comets lose their ice and dust and become rocky asteroidal hulks. An asteroid-type remnant of an exhausted comet can no longer form a coma or tail. It can only be seen as a dark rock reflecting a tiny bit of the Sun's light.

The dust that comets release continues in orbit within the solar system and is probably responsible for zodiacal light. The zodiacal light is best seen in the west after evening twilight in the winter and spring or in the east before morning twilight in the summer and autumn. It is sometimes called "false dawn" because its diffuse and subtle light, centered on the ecliptic plane, can be mistaken for the Sun's early twilight glow. The zodiacal light is due to the Sun's rays reflected from the tiny dust particles of ancient comets.

High-flying aircraft, equipped with special collection devices, may have actually captured some of this fragile dust. Though cometary dust rains down upon Earth's atmosphere on a daily basis, it is unlikely that any sizeable pieces have survived the fiery trip through the Earth's atmosphere. Sizeable meteorites that have survived the atmospheric heating to land on the Earth's surface are thought to be chunks of asteroids, not comets.

journey sunward. Over a series of visits to the planetary realm of the solar system, first Uranus and Neptune shortened its orbit until finally Jupiter deflected it so profoundly that never since has the comet ventured beyond the orbit of the outermost planet.

For centuries this comet came and went, leaving Earthlings first panic-stricken, then fascinated and appreciative. It was not always bright, but it was a dependable visitor; its faithfulness linked generations across a lifespan. It is the only short-period comet that can be regularly seen with the eyes alone. The people of Earth named it Halley's Comet.

Sometime in the future, Halley's Comet will be seen no more—the ultimate fate of comets. It may exhaust all its gas-producing ices and become only a dark, rocky, asteroidlike hulk no longer able to produce a coma and a tail. Or Halley's Comet may break up into countless bits of debris. In either case, left behind in orbit around the Sun, still traveling along and near the path the comet once took, will be a stream of tiny dust particles that once helped to form the comet's awesome coma and tail.

With each visit to the Sun, Halley's Comet sheds material. Each year, in early May and late October, as the Earth passes near the orbit of Comet Halley, we collide with these fragments at a speed of 40 miles (65 kilometers) per second.[4] The friction of the air heats these particles to incandescence, and they burn up 50 miles (80 kilometers) above our heads to produce meteors or "shooting stars." Most meteor particles are the size of a grain of sand.

One day, two meteor showers may be all that remains of Halley's Comet. But for now, the comet has returned and is ours to appreciate and remember for our lifetime.

In anticipation of the 1986 apparition of the most famous of comets, noted anthropologist and science writer Loren Eiseley recorded his thoughts:

> In the year 1910 Halley's Comet...was again brightening the night skies of earth....
>
> Like hundreds of other little boys of the new century, I was held up in my father's arms...to see the hurtling emissary of the void. My father told me some-

thing then that is one of my earliest and most cherished memories.

"If you live to be an old man," he said carefully, fixing my eyes on the midnight spectacle, "you will see it again. It will come back in [seventy-six] years. Remember," he whispered in my ear, "I will be gone, but you will see it. All that time it will be traveling in the dark, but somewhere, far out there"—he swept his hand toward the blue horizon of the plains—"it will turn back...."

I tightened my hold on my father's neck and stared uncomprehendingly at the heavens. Once more he spoke against my ear and for us two alone. "Remember, all you have to do is be careful and wait.... I think you will live to see it—for me...."

"Yes, Papa," I said dutifully. [His] voice still sounds in my ears and I know with the sureness of maturity that the great wild satellite has traversed its course and is speeding on its homeward journey toward the Sun.[5]

Halley's Comet has returned. It will not come our way again until the year 2061.

APPENDICES

Appendix 1
Profile of Halley's Comet

Physical Data

Diameter (estimated)	3.7 miles (6 kilometers)
Density (estimated)	1 gram per cubic centimeter (density of water)
Mass (estimated)	100 billion tons (10^{17} grams)
Composition (based on typical comets)	Ices of water (H_2O), carbon monoxide (CO), carbon dioxide (CO_2), ammonia (NH_3), methane (CH_4), cyanide compounds (HCN & CH_3CN), and others, plus solid particle compounds of carbon, silicon, sodium, potassium, calcium, iron, sulfur, and others
Rotation on axis	Approximately 53 hours, direct (spins same direction as its orbital motion)
Revolution around Sun	Averages 76.1 years
	shortest period: 74.42 years (1835–1910)
	longest period: 79.25 years (A.D. 451–530)
Orbital characteristics	Comet motion around Sun is retrograde (opposite direction from planets)
	Orbit is inclined to plane of Earth's orbit by 18° (or 162°, depending on which angle you measure)
Escape velocity (estimated)	4.5 miles per hour (2 meters per second)

Historical Data

Earliest recorded appearance (reasonably certain)	240 B.C.
Number of recorded apparitions from 240 B.C. to A.D. 1986	30
Closest known approach to Earth	3.1 million miles (4.9 million kilometers) on April 10, A.D. 837
Greatest known brightness	Approximately magnitude –3.5 (ca. half the brightness of Venus) on April 10, A.D. 837
Longest tail recorded (in angular length)	150° on May 20, 1910

1986 Return

Perihelion (closest to Sun)	February 9, 1986: 54.5 million miles (87.8 million kilometers; 0.587 astronomical units)
Perihelion velocity	33.8 miles per second (54.6 kilometers per second)
Aphelion (farthest from the Sun)	1948, 2023: 3.28 billion miles (5.28 billion kilometers; 35.28 astronomical units)
Aphelion velocity	0.6 miles per second (0.9 kilometers per second)
Closest approach to Earth	Inbound—November 27, 1985: 57.6 million miles (92.8 million kilometers) Outbound—April 10, 1986—39.0 million miles (62.8 million kilometers)

Appendix 2
Halley's Comet Through History

Perihelion Date	What Was Happening in Europe and America
May 25, 240 B.C.	Aristarchus proposes that the Earth and other planets orbit the Sun (ca. 250 B.C.); Romans destroy Carthaginian fleet off Sicily, and Sicily becomes a Roman province (241 B.C.); Eratosthenes accurately calculates the circumference of Earth (ca. 230 B.C.).
November 12, 164 B.C.	Jews win religious freedom from Greco-Syrians (165 B.C.); Hipparchus classifies stars by brightness and creates first major star catalog (ca. 160 B.C.).
August 6, 87 B.C.	Civil war in Rome is fought (88 B.C.).
October 10, 12 B.C.	Rome expands its empire in Europe.
January 25, A.D. 66	Jews revolt against Rome (A.D. 66); Rome destroys Jerusalem (A.D. 70).
March 22, 141	Rome, after crushing a new Jewish revolt (135), quashes a rebellion in Britain (143); Ptolemy codifies prevalent Greek idea that the Earth is the center of the universe with the Sun and planets orbiting it in circles.
May 17, 218	Goths invade Asia Minor and southeastern Europe (ca. 220).
April 20, 295	Diocletian divides Roman Empire into eastern and western parts (285).
February 16, 374	Huns invade eastern Europe; books begin to replace scrolls (ca. 360).
June 28, 451	Huns under Attila invade France (Gaul) and are defeated at Châlons (451).
September 27, 530	Efforts are made to reunite the Christian Roman Empire.
March 15, 607	Mathematicians in India develop decimal position (by 604); Muhammad founds Islam (612).
October 2, 684	Moslem siege of Constantinople fails (678).
May 20, 760	Charlemagne becomes sole king of Franks (771).
February 28, 837	Moslems suppress a revolt of Jews and Christians in Toledo, Spain (837).

July 18, 912	Moslem rule in Spain begins its golden era.
September 5, 989	Viking Eric the Red discovers and colonizes Greenland (982).
March 20, 1066	William the Conqueror invades England and wins the Battle of Hastings (October 14, 1066).
April 18, 1145	Pope calls for Second Crusade.
September 28, 1222	Fifth Crusade ends in failure (1221).
October 25, 1301	French parliament meets for first time (1301); Giotto paints "The Adoration of the Magi" with a comet over the manger (ca. 1304).
November 10, 1378	"Great Schism" begins in Roman Catholic Church (1378).
June 9, 1456	War of the Roses begins in England (1455); János Hunyadi (Hungary) repels Ottoman Turks at Belgrade but dies of the plague (1456).
August 26, 1531	Peter Apian is first European to illustrate that comet tails always flow away from Sun (1531); Copernicus circulates his manuscript describing the Earth and planets orbiting the Sun; Pizarro sets off to conquer the Incas in Peru.
October 27, 1607	Jamestown, first permanent English settlement on the American mainland, is established (1607); forks for eating come into use in England and France, having originated in Italy (1607); Shakespeare writes *Antony and Cleopatra* (1606); Kepler publishes concept that planets move in elliptical orbits around Sun (1609); Galileo makes first astronomical telescope (1609).
September 15, 1682	La Salle claims Louisiana Territory for French; Edmond Halley sees comet that will one day carry his name.
March 13, 1759	The comet seen by Halley returns as he predicted and is soon called Halley's Comet; English are winning French and Indian War in North America; Voltaire publishes *Candide*; Haydn writes the first of his 104 symphonies.
November 16, 1835	Mark Twain is born; Texans revolt against Mexican rule; Samuel Colt patents his revolver; Hans Christian Andersen publishes his first volume of fairy tales; P. T. Barnum begins his show business–circus career.

April 20, 1910   Tail of Halley's Comet grazes Earth; Marie Curie isolates radium; "The Firebird" ballet by Stravinsky premieres; Mark Twain dies.

NOTE: Information taken from the following sources:

Bernard Grun. *The Timetables of History*. New York: Simon and Schuster, 1979.

Queensbury Group. *The Book of Facts*. New York: Paddington Press, 1978.

Appendix 3
Comet Halley and Comet Giacobini–Zinner Studies from Spacecraft

Spacecraft Name (Country)	Launch Date	Arrival Date	Flyby Distance (km)
Halley Missions Pioneer (Japan)	1/7/85	3/11/86	7 million sunward
Planet A (Japan)	8/20/85	3/8/86	200,000 sunward
Giotto (ESA)[a]	7/2/85	3/14/86	500 sunward
Vega 1 (Soviet Union)[b]	12/15/84	3/6/86	10,000 sunward
Vega 2 (Soviet Union)[b]	12/21/84	3/9/86	less than 10,000 sunward
Giacobini–Zinner Mission ICE (USA)[c]	12/22/83	9/11/85	10,000 tailward

[a] ESA's member countries include Belgium, Denmark, France, Germany, Great Britain, Ireland, Italy, Netherlands, Spain, Sweden, and Switzerland.

[b] The Vega project includes major contributions from the USSR, Hungary, France, and even the United States. John A. Simpson of the University of Chicago built the dust detectors for the Vegas. The work was announced only after both Vegas had been launched. Each Vega spacecraft will fly by the planet Venus in June 1985 where it will drop off both an atmospheric balloon and a surface lander before proceeding on to its encounter with Comet Halley.

Flyby Velocity (km/s)	Mission Goals and Science Instrumentation
70	Test of launch vehicle, spacecraft, and deep space communications. Three science experiments: solar proton and plasma wave analyzers and magnetometer (used to study solar wind conditions near Halley).
70	Study Halley's outer atmosphere. Two science experiments: solar wind spectrometer used to study distribution of solar ions and electrons and ultraviolet spectrometer used to measure giant hydrogen cloud surrounding comet's head.
68	Measure gas and dust environment and image nucleus. Ten science experiments. Imaging camera and photopolarimeter will do remote sensing of comet. Closer to the comet's nucleus, the comet's gas and dust will be studied with a dust impact detector and three spectrometers (neutral mass, ion mass, and dust mass). The comet's charged-particle environment will be examined by a magnetometer and three charged-particle detectors.
78 78	Measure gas and dust environment and image nucleus. Vega 1 and 2 are identical spacecraft with twelve science experiments each. With the exception of the polarimeter, the Vega science experiments are similar to those on Giotto. In addition, the Vegas include another charged-particle instrument (an infrared sounder) and both narrow and wide-angle imaging cameras.
21	Study charged-particle environment of comet's tail. Although the original spacecraft had thirteen on-board science experiments, only six working instruments are suitable for studying the comet's tail environment.

[c]The International Cometary Explorer (ICE) spacecraft was originally placed in orbit around the Sun on August 12, 1978, to monitor the solar wind. In June 1982, a complex set of spacecraft maneuvers was initiated to put the probe into an Earth orbit and then, by using a very close lunar swingby on December 22, 1983, to send the spacecraft on a trajectory that would allow it to intercept Comet Giacobini–Zinner on September 11, 1985. Although NASA has operational responsibility for the ICE spacecraft, the science team includes experimenters from England, France, and West Germany as well as the United States.

Appendix 4
Major Meteor Showers and Their Associated Comets

Shower Name	Shower Dates	Current Shower Maximum	Maximum Expected Hourly Rate
Quadrantid	Jan. 1–6	Jan. 4	40
Lyrid	Apr. 19–24	Apr. 22	12
Eta Aquarid	May 1–8	May 5	20
Southern Delta Aquarid	July 15–Aug. 15	July 27	35
Perseid	July 25–Aug. 18	Aug. 12.5	68
Giacobinid	Oct. 8–10	Oct. 8	—
Orionid	Oct. 16–26	Oct. 20	30
Leonid	Nov. 15–19	Nov. 17	—
Geminid	Dec. 7–15	Dec. 14.4	58

[a]"P/" means short-period comet.

[b]Possible, but not too likely, Giacobinid meteor shower on Oct. 8.6, 1985.

[c]Possible Leonid meteor shower or storm on Nov. 17.4, 1997, and probable meteor shower or storm on Nov. 17.5, 1998, or Nov. 17.8, 1999.

Associated Comet[a]	Constellation of Radiant	Comments
—	Bootes	annual showers
1861 I	Lyra	annual showers
P/Halley	Aquarius	annual showers
—	Aquarius	annual showers
P/Swift–Tuttle	Perseus	annual showers
P/Giacobini–Zinner	Draco	periodic storms[b]
P/Halley	Orion	annual showers
P/Tempel–Tuttle	Leo	periodic storms[c]
—	Gemini	annual showers

Appendix 5
Comets Observed to Have Split

Comet[a]	Numerical Designation	Observed No. of Pieces	Comet Split[b]	Computed Date of Split(s)	Distance from Sun[c]
P/Biela	1846 II	2	A	May 25, 1840	3.59
Liais	1860 I	2	B	Sept. 18, 1859	2.49
Great Sept. Comet	1882 II	4	A	Sept. 17, 1882	0.017
Sawerthal	1888 I	2	B	Mar. 2, 1888	0.76
Davidson	1889 IV	2	A	Jul. 30, 1889	1.06
P/Brooks 2	1889 V	5	B	Jul. 21, 1886	5.38
			B	Feb. 10, 1888	4.25
P/Giacobini	1896 V	2	B	Apr. 24, 1896	2.36
Swift	1899 I	3	A	Apr. 25, 1899	0.48
			A	May 28, 1899	1.15
Kopff	1905 IV	2	A	Dec. 11, 1905	3.38
Campbell	1914 IV	2	A	Aug. 25, 1914	0.82
Mellish	1915 II	5	B	Feb. 26, 1915	2.38
			B	Mar. 23, 1915	2.09
P/Taylor	1916 I	2	B	Dec. 8, 1915	1.65
Whipple– Fedtke– Tevzadze	1943 I	2	A	Mar. 9, 1943	1.43
Southern Comet	1947 XII	2	B	Nov. 30, 1947	0.15
Honda	1955 V	2	B	July 1953	8.2
Wirtanen	1957 VI	2	B	Sept. 10, 1954	9.25
Ikeya–Seki	1965 VIII	2	A	Oct. 21, 1965	0.008
Wild	1968 III	2	A	Aug. 3, 1968	2.92
Tago–Sato– Kosaka	1969 IX	2	A	Feb. 9, 1970	1.20
Kohoutek	1970 III	2	A	Apr. 29, 1970	1.79
West	1976 VI	4	B	Feb. 19, 1976	0.30
			A	Feb. 27, 1976	0.22
			A	Mar. 6, 1976	0.41

[a]"P/" means short-period comet.
[b]Key: B, before perihelion; and A, after perihelion.
[c]When the split occurred, in astronomical units.

NOTE: Table is adapted from the work of Zdenek Sekanina.

NOTES

Prologue and Chapter 1

[1] At the time of its "recovery," Halley's Comet had a magnitude of 24.2. The comet was within eight arc seconds of the position predicted by Yeomans.

[2] Paraphrased from Aristotle, especially *Meteorology*, Chapters 6 and 7, on comets.

[3] Fred L. Whipple notes that the Greeks got the idea of designating such objects as "hairy stars" from the Egyptians.

[4] The adjective "Holy" to describe the Roman Empire was not employed until the latter half of the twelfth century.

The quotation is from (author anonymous) *Son of Charlemagne: A Contemporary Life of Louis the Pious*, translated by Allen Cabaniss (Syracuse: Syracuse University Press, 1961), pp. 112–113.

[5] Also novae (exploding stars), which the Chinese called "guest stars."

[6] Joseph Needham, *Science and Civilization in China* (Cambridge, England: Cambridge University Press, 1959), Vol. 3, p. 432.

[7] Earlier victories in France against other dukes had already started people referring to William as "the Conqueror."

[8] The Bayeux Tapestry is actually an embroidery. Its pictures were created by sewing, not weaving. But this work has been called the Bayeux Tapestry for so long that the name has stuck.

[9] The often-told story that the Pope excommunicated or issued a bull against the comet is not true.

[10] Apian (1495–1552) latinized his German name of Peter Bienewitz to Petrus Apianus for his published work. His latinized name is frequently anglicized as Peter Apian.

Fracastoro (ca. 1478–1553) is better known as the physician whose idea that diseases are spread by invisible living particles provided the framework for the modern understanding that germs are agents of disease.

It is doubtful that Seneca, the Roman philosopher (4 B.C.–A.D. 65), truly foreshadowed both Apian and the Chinese when he wrote in his *Naturales Questiones* (Natural Questions), Book VII (Comets), 20:4: "...the tails of comets fly from the sun's rays." The reference is ambiguous because Seneca is discussing comets that are not visible because of their closeness to the Sun. The full sentence reads, "Obviously the comet itself is blanketed by the light of the sun and so cannot be seen, but the tail escapes the sun's rays." (Translation by Thomas H. Corcoran [London: William Heinemann, 1972].) Later (Book VII,

26:2), Seneca inquires into the elongated shape of comets and mentions the rays of the Sun, but does not relate the rays to the direction the tail extends.

[11] Some scholars say it was copper.

[12] Andreas Celichius: *Theologische erinnerung, von dem newen Cometen* (Magdeburg: Joachim Walden, 1578).

[13] Tycho measured the parallax at fifteen minutes of arc or less; therefore, the comet had to be at least 240 Earth radii distant.

[14] Tycho observed a supernova (giant exploding star) in 1572, and it showed no parallax at all. Clearly it was as distant as the stars, and yet it was changing. This observation, which was made five years before the Comet of 1577, provided evidence that the universe was not as the Greeks and the medieval Christian Church believed.

[15] "They are not eternal, as Seneca imagined.... The direct rays of the Sun strike upon it, penetrate its substance, draw away with them a portion of this matter, and issue thence to form the track of light we call the tail of a comet. This action of the solar rays attenuates the particles which compose the body of the comet. It drives them away; it dissipates them. In this manner the comet is consumed by breathing out, so to speak, its own tail." (Cited by Charles P. Olivier: *Comets* [Baltimore: Williams & Wilkins, 1930], pp. 9–10.)

It wasn't until the middle of the twentieth century that we understood that particles from the Sun and the pressure of sunlight are both responsible for the tails of comets.

[16] Guillaume du Bartas: *Du Bartas His Divine Weekes and Workes*, translated by Josuah Sylvester (London, 1605).

Chapter 2

[1] Coffeehouses were the rage at that time. Many also sold that popular but expensive new beverage from Central America called hot chocolate.

[2] So said Halley in a letter to Newton. Isaac Newton: *The Correspondence of Isaac Newton*, Vol. 2 (1676–1687), edited by H. W. Turnbull (Cambridge, England: Cambridge University Press, 1960), p. 442.

[3] Newton called this mathematics "fluxions." Gottfried Wilhelm Leibniz (1646–1716) independently invented this same tool a few years later and called it "calculus." The notation Leibniz employed is easier to use, and it is in this form that calculus is taught today.

⁴ The story of the first meeting between Halley and Newton was recounted anecdotally much later by an acquaintance of Newton or Halley named Abraham de Moivre in a letter to John Conduitt. De Moivre was not present at the meeting and presumably heard it recalled by one or both of the great scientists. It is reported in most studies of Newton, such as I. Bernard Cohen: *Introduction to Newton's "Principia"* (Cambridge, Massachusetts: Harvard University Press, 1971), p. 50.

The simulated dialogue is based on de Moivre's letter.

⁵ It was November when Edward Paget, a mathematics teacher, brought Halley the proof from Newton (Colin A. Ronan: *Sir Isaac Newton* [London, International Textbook, 1969], p. 39). Newton also decided to present for the first time his concept of gravity to his students that fall in a series of nine lectures entitled *De Motu Corporum* (On the Motion of Bodies). The course was a flop. The students couldn't understand Newton's lectures. The course was not offered again.

⁶ Newton dealt with comets to a limited extent in the first edition of *Principia*, providing a means of computing parabolic (nonreturning) orbit approximations for comets based on three good-quality observations. Halley furnished the parabolic orbit diagrams for the *Principia* based on observations of the Comet of 1680. Newton then went on to add in his third and final (1726) edition of *Principia*:

> The orbit is determined... by the computation of Dr. Halley, in an
> ellipse. And it is shown that... the comet took its course through the
> nine signs of the heavens, with as much accuracy as the planets move
> in the elliptic orbits given in astronomy.... [Thus] comets are a sort
> of planet revolved in very eccentric orbits around the sun.

(This translation from Book III of *Principia* is by Andrew Motte [1729]).

⁷ Seneca, in the first century A.D., believed that comets travel along planetlike paths. "Why, then, are we surprised that comets, such a rare spectacle in the universe, are not yet grasped by fixed laws and that their beginning and end are not known, when their return is at vast intervals?" (*Naturales Questiones*, Book VII, 25:3; translated by Thomas H. Corcoran). Seth Ward in England suggested in 1653 that comets have a "circular or elliptical orbit." However, Ward's views were based not on observational evidence but rather on his belief that comets were eternal and hence must move in closed orbits.

⁸ From Halley's *Astronomiae Cometicae Synopsis (Synopsis of the Astronomy of Comets)* (Oxford & London: John Senex, 1705), p. 22. The version given

here is a free translation. The 1705 English version of *Synopsis*, "translated from the Latin," reads:

> Hence I dare venture to foretell, that it will return again in the year 1758. And, if it should then return, we shall have no reason to doubt but the rest must return too.

In this brief volume, Halley also identified the Comet of 1456 as a probable appearance of the comet we now know by his name. It had not been carefully observed by Europeans, "yet from its period and manner of its transit," Halley wrote, "I cannot think [it] different from those" that Apian (1531), Kepler (1607), and he (1682) had seen (p. 22).

In a later edition of *Synopsis*, Halley also erroneously identified the comets of 1305 and 1380 as early apparitions of the returning comet. He then went on to add:

> ...wherefore if according to what we have already said it should return again about the year 1758, candid posterity will not refuse to acknowledge that this was first discovered by an Englishman. (Angus Armitage: *Edmond Halley* [London, Thomas Nelson and Sons, 1966], p. 166.)

[9] Halley's Comet reached perihelion on March 13, 1759. Halley (in *Synopsis*, 1705, pp. 21–22) and succeeding astronomers understood that gravitational perturbations by the planets, especially Jupiter and Saturn, could change comet orbits and hence their orbital periods.

[10] All of the comet apparitions mentioned in Chapter 1 were returns of Halley's Comet *except* for the Comet of 1577 which Tycho Brahe observed so well that it became a turning point in mankind's understanding of comets.

[11] Until recently, no historical document had been found which recorded Comet Halley's visit to the Sun in 164 B.C. However, the science historians Richard Stephenson, Herman Hunger, and Kevin Yau have located in a Babylonian text a sighting of a comet in 164 B.C. The time of year and the comet's motion clearly identify it as Halley's.

Chapter 3

[1] Some comets will leave our solar systems because of gravitational perturbations by planets, especially Jupiter. All observed comets seem to have originated in the solar system, but this was not known to Halley.

[2] Halley's Comet has a "retrograde" orbit; that is, it travels around the Sun in the opposite direction of the Earth and all the planets. About half of the long-

period comets have retrograde orbits. Halley's Comet is one of the very few short-period comets with a retrograde path. The giant planets, especially Jupiter, gravitationally perturb long-period comets into shorter orbits and tend to disturb them into paths that carry the comets around the Sun in the same direction the planets travel. The passage of a comparatively small celestial body across the face of the Sun is called a "transit."

[3] The passage of the Earth through the tail of Tebbutt's Comet on June 30, 1861, was said to have caused a yellow auroralike glow in the sky and to have briefly dimmed the Sun (Peter Lancaster Brown: *Comets, Meteorites, and Men* [New York: Taplinger, 1974], p. 84), but no specific chemical or physical effects were measured.

[4] "Disaster" comes from Latin and means "evil star."

[5] *The New York Times*, May 18, 1910, p. 1.

[6] *The Washington Post*, May 23, 1910, p. 1.

[7] Bill Stephenson: "The Panic Over Halley's Comet," *Maclean's*, May 14, 1955, pp. 30–31+. This article contains so many scientific errors that the accuracy of its stories is also suspect.

[8] Dinsmore Alter: "Comets and People," *Griffith Observer*, July 1956, p. 82.

[9] The Space Shuttle's main engines burn liquid hydrogen and liquid oxygen, pound for pound the most powerful chemical propellant mixture. The exhaust of these three rocket motors is water.

[10] Richard A. Proctor, as quoted by D. J. McAdam (Professor of Astronomy at Washington and Jefferson College): "The Menace in the Skies: The Case for the Comet," *Harper's Weekly*, Vol. 54, May 14, 1910, p. 12.

[11] Jerry Klein: "When Halley's Comet Bemused the World," *New York Times Magazine*, May 8, 1960, pp. 45+.

[12] Albert Bigelow Paine: *Mark Twain: A Biography* (New York: Harper & Brothers, 1912), Vol. 3, p. 1511.

[13] April 20 Universal Time (Greenwich Mean Time); April 19 in the United States.

[14] The Earth passed through the ion (gas or plasma) tail. Plasma in physics and astronomy means a very hot gas made up of electrons and positively charged (ionized) particles. The dust tail of Halley's Comet did not sweep across the Earth.

Chapter 4

[1] Quoted by Greg Stone: "'A Sort of Heavenly Polywog'," *Yankee*, Vol. 37, No. 12 (December 1973), p. 93.

[2] At aphelion, Halley's Comet is 9.99 astronomical units below the plane of the ecliptic.

[3] Uranus has fourteen times the Earth's mass and a diameter of 32,200 miles (51,900 kilometers). Neptune, somewhat denser, packs seventeen times the mass of the Earth into a diameter of 31,000 miles (50,000 kilometers). As of 1984, Uranus was known to have five moons; Neptune was known to have two.

[4] The hydrogen cloud that surrounds a comet's coma was discovered by NASA's Orbiting Astronomical Observatory 2 (OAO–2) in January 1970 while examining Comet Tago–Sato–Kosaka (1969 IX) with a telescope sensitive to ultraviolet wavelengths that do not pass through the Earth's atmosphere. The cloud was bigger than the Sun. Later that year, NASA's Orbiting Geophysical Observatory 5 (OGO–5) confirmed the hydrogen cloud with ultraviolet observations of Comet Bennett (1970 II). The cloud around Bennett was eight times the diameter of the Sun.

[5] Venus also has craters, but we cannot see them through its cloudy atmosphere. Their existence is known from radar signals bounced off Venus from the Earth.

[6] Antonio Luzcano-Araujo and J. Oró: "Cometary Material and the Origins of Life on Earth," Cyril Ponnamperuma (editor): *Comets and the Origin of Life* (Boston: D. Reidel, 1981), p. 192. See also A. H. Delsemme: "Are Comets Connected to the Origin of Life?" in the same book.

Phosphorus is probably present in comets but is hard to detect because its expected abundance is very low.

[7] J. John Sepkoski, Jr. in a personal communication. Dale A. Russell says that more than seventy-five percent of all plant and animal species perished in "The Mass Extinctions of the Late Mesozoic," *Scientific American*, January 1982, pp. 58–65.

[8] David A. Raup and J. John Sepkoski, Jr.: "Periodicity of Extinctions in the Geologic Past," *Proceedings of the National Academy of Sciences USA*, Vol. 81, February 1984, pp. 801–805.

[9] Luis W. Alvarez, Walter Alvarez, Frank Asaro, and Helen V. Michel: "Extraterrestrial Cause for the Cretaceous–Tertiary Extinction," *Science*, Vol. 208, 6 June 1980, pp. 1095–1108.

[10] The impacting object could have shattered in the air or smashed into land or water. If it hit land, the crater and crustal deformation should still be visible.

[11] Raup and Sepkoski, op. cit.

[12] Walter Alvarez and Richard A. Muller: "Evidence from Crater Ages for Periodic Impacts on the Earth," *Nature*, Vol. 308, 19 April 1984, pp. 718–720. See also Zdenek Sekanina and Donald K. Yeomans: "Close Encounters and Collisions of Comets with the Earth," *Astronomical Journal*, Vol. 89, No. 1 (January 1984), pp. 154–161.

[13] Information provided by Barringer Meteor Crater staff.

[14] Daniel P. Whitmire and Albert A. Jackson IV: "Are Periodic Mass Extinctions Driven by a Distant Solar Companion?" (pp. 713–715) and Marc Davis, Piet Hut, and Richard A. Muller: "Extinctions of Species by Periodic Comet Showers" (pp. 715–717) *Nature*, Vol. 308, 19 April 1984.

[15] Siva is pronounced "SHE-va." Are we ready for a science fiction story about a future approach of our Sun's companion entitled "Come Back, Little Siva"?

[16] Michael R. Rampino and Richard B. Stothers: "Terrestrial Mass Extinctions, Cometary Impacts and the Sun's Motion Perpendicular to the Galactic Plane," *Nature*, Vol. 308, 19 April 1984, pp. 709–712.

[17] The comet storm scenario is based on work of J. G. Hills and is cited by the authors of both hypotheses. See "Comet Showers and the Steady-State Infall of Comets from the Oort Cloud," *Astronomical Journal*, Vol. 86, No. 11 (November 1981), pp. 1730–1740.

[18] Stephen J. Gould: "The Cosmic Dance of Siva," *Natural History*, August 1984, pp. 14–19, and "The Ediacaran Experiment," *Natural History*, February 1984, pp. 14–23.

[19] On each of these dates, the comet is "crossing" the planet's orbit in the sense that an observer on the planet's orbit would look perpendicularly to the ecliptic (the plane of the solar system) to see the comet. Because the orbit of Halley's Comet is inclined to the ecliptic by 17.75 degrees, when the comet "crosses" a planet's orbit, it is farther from the Sun than that planet.

[20] For a list of comets that have been observed to split while passing the Sun, see Appendix 5.

A comet need not break up or vaporize entirely. It may fade away into inactivity with its nucleus no longer able to form a coma and tail. Some or most of the Apollo asteroids that pass close to the Earth have been hypothesized to be the inactive nuclei of comets rather than standard asteroids.

[21] Halley's Comet has a mass of approximately 100 billion tons ($10^{17\pm1}$ grams).

[22] *National Geographic* news release on a comet visible only by telescope, March 31, 1955.

Chapter 5

[1] Sakigate was originally designated MS–T5.

[2] Some American scientists are coinvestigators on experiments aboard the ESA Giotto spacecraft and the Soviet Vegas. The Vegas also carry American-built dust analyzers.

[3] ISEE–3 was in orbit around the point between the Earth and Sun where its position was nearly stable with respect to small perturbing forces. This position is known as a Langrangian point, after the mathematician who discovered that in a three-body system there will be five such points where a small body can remain nearly in the same position with respect to the two larger co-orbiting objects. ISEE–3 orbited the first of these positions—"L1."

[4] During this orbital adjustment phase, the spaceprobe twice looped far into the region "down-Sun" from Earth. There it sampled the cometlike "tail" of the Earth which results from the interaction of the solar wind and the charged gases (plasma) of the Earth's upper atmosphere (ionosphere).

Chapter 6

[1] The Jet Propulsion Laboratory is operated for NASA by the California Institute of Technology.

Eastern Hemisphere headquarters for the International Halley Watch (IHW) is Bamberg, West Germany. The two leaders for the IHW are Ray L. Newburn, Jr. and Jürgen Rahe.

[2] The period of Comet Ikeya–Seki (1965 VIII) is approximately 880 years, and that of Bennett (1970 II) is 1700 years. Brian G. Marsden thinks it is probable that Comet West (1976 VI) has been ejected from our solar system and will never return.

[3] Approximately the brightness of Comet IRAS–Araki–Alcock (1983d) in May 1983. A comet of second magnitude does not appear as bright as a star of second magnitude because all the star's brightness is concentrated into a point of light, whereas a comet's total brightness is spread over its coma and tail.

John E. Bortle and Charles S. Morris predict that the central region of the coma of Halley's Comet may be close to Polaris in brightness ("Brighter Prospects for Halley's Comet," *Sky and Telescope*, January 1984, pp. 9–12). Other astronomers, however, predict that Halley's Comet will reach only one-third to one-half the brightness of Polaris.

[4] At perihelion, Halley's Comet will be visually unobservable from Earth, but radio and infrared observations will be attempted. It will also be monitored by

two American spacecraft: the Solar Maximum Mission satellite in Earth orbit (repaired by Space Shuttle astronauts in 1984) and the Pioneer Venus Orbiter circling Venus.

Chapter 7

[1] Of course, Jupiter's gravity can also fling comets out of the solar system. That certainly happened to millions of comets since the solar system began.

[2] A similar idea was proposed in 1932 by the Estonian astronomer Ernst J. Opik.

[3] J. G. Hills and M. T. Sandford II: "The Formation of Comets by Radiation Pressure in the Outer Protosun," *Astronomical Journal*, Vol. 88, No. 10 (October 1983), pp. 1519–1530.

[4] The May meteor display is known as the Eta Aquarid Shower; the October display is called the Orionid Shower. These designations refer to Aquarius and Orion, the constellations from which these meteors seem to radiate. The radiant in Aquarius is further specified by the star Eta Aquarii to differentiate it from other meteor showers that have meteors whose paths diverge from Aquarius.

[5] From "The Star Dragon" in *The Invisible Pyramid* (New York: Charles Scribner's Sons, 1970).

Loren Eiseley lived from 1907 to 1977.

ACKNOWLEDGMENTS

We both would like to thank Frank Bigger, Manager, Public Understanding of Science Projects, American Chemical Society. This project began and became reality because of his imagination and dedication. So capably assisting him and us were Laverne R. Cochrell and Vicki A. Hickman.

Valuable technical and aesthetic help with this manuscript and its illustrations was provided by Frank Bigger, John C. Brandt, Ruth S. Freitag, Brian G. Marsden, and Harry L. Shipman.

Thanks to Vincent R. Tocci, Head, Department of Public Communication, American Chemical Society, for his encouragement and support.

We are grateful to the American Chemical Society and its Books Department: M. Joan Comstock, Head; Robert H. Johnson, Acquisitions Editor; Karen McCeney, Editorial Assistant; Pamela Lewis, Artist; Hilary Kanter, Production Assistant; and Kathie Friedley, Marketing Associate.

Each of us has individual thanks to express as well.

Don would like to thank his family and his father George E. Yeomans for their support and Philippe Veron for sharing his interest in cometary history and foolishness.

For their ideas and encouragement, Mark would like to thank his wife, Peggy, his parents, Lewis and Muriel Littmann, David and Esther Littmann, Carl Littmann, Ann and Paul Rappoport, Jane Littmann, and Bea and Tom Owens. Bea Owens was also the tireless manuscript typist.

This book grew from the continuing effort of the American Chemical Society to promote understanding and appreciation for the sciences—an effort that began formally in 1919. The basic information for this book was originally contained in an American Chemical Society planetarium program made available free to planetariums throughout the world. That program, also called "Comet Halley: Once in a Lifetime," was made possible by grants to the American Chemical Society by the Bushnell Optical Division of Bausch & Lomb, Inc., the American Association for the Advancement of Science, and the American Astronomical Society. These organizations are long-standing friends of planetariums and science education. We would specifically like to thank Mark R. O'Brien, Bausch & Lomb; William D. Carey, American Association for the Advancement of Science; and Harry L. Shipman, American Astronomical Society. They were instrumental in obtaining the seminal grants to the American Chemical Society for the production and distribution of the planetarium program.

Serving most effectively as technical advisers for the planetarium program (and thus for the formative stages of this book) were Michael J. S. Belton (Kitt Peak National Observatory), John C. Brandt (NASA Goddard Space Flight Center), Ruth S. Freitag (Library of Congress), Jane P. Geoghegan (Thomas Jefferson High School Planetarium, Richmond, Virginia), Richard M. Lemmon (University of California, Berkeley), Brian G. Marsden (Smithsonian Astrophysical Observatory), Ray L. Newburn, Jr. (International Halley Watch), Harry L. Shipman (University of Delaware), Fred L. Whipple (Smithsonian Astrophysical Observatory), and Donald K. Yeomans (Jet Propulsion Laboratory).

Thanks also to Charles D. Smith, Director, Universe Planetarium/Space Theater, Science Museum of Virginia (Richmond), and to his staff, producers of the planetarium program, for their help with the script.

For their assistance with special sections of this book, we are grateful to Robert W. Farquhar (NASA Goddard Space Flight Center), Daniel W. E. Green (Smithsonian Astrophysical Observatory), John G. Hills (Los Alamos National Laboratory), Dorrit Hoffleit (Yale University), Owen Gingerich (Harvard University), J. John Sepkoski, Jr. (University of Chicago), and Richard W. Shorthill (University of Utah).

GLOSSARY

absorption (dark-line) spectrum   The spectrum provided by a glowing object when seen through a cooler gas. The cooler gas absorbs specific wavelengths on the basis of the chemicals present, and the resulting spectrum exhibits a series of a dark lines from which the composition and temperature of the intervening gas can be determined.

altitude   In astronomy, an angular measurement of position from the horizon upward. The zenith (the point overhead) has an altitude of ninety degrees from a level horizon.

angstrom   A unit of measure for the wavelength of light: one angstrom is one ten-millionth (10^{-8}) of one centimeter, approximately the diameter of a hydrogen atom.

antitail (of a comet)   A comet tail that appears to extend toward the Sun. It is visible when the Earth passes through the orbital plane of a comet. When this situation occurs, dust particles in the comet's orbital plane are seen projected on the starfield as a sunward tail-like feature.

aphelion   The point in an object's orbit around the Sun at which it is farthest from the Sun. (At this point, the object is traveling at its slowest speed.)

apparition   The appearance of a comet (or other celestial object) after it has been out of sight for a length of time.

asteroid   A planetlike body in the solar system too small to be classified as a planet. Most asteroids (or minor planets) orbit the Sun between Mars and Jupiter.

astronomical unit   A unit of measure in astronomy approximately equal to the average distance between the Earth and Sun: 92,960,000 miles (149.6 million kilometers). Jupiter, for example, is 5.2 astronomical units from the Sun.

azimuth   An angular measurement of position along the horizon, usually starting from north and moving clockwise. East is azimuth ninety degrees.

binary star system   Two stars revolving around one another because of gravity.

broadside   In journalism, in earlier times, a large sheet of paper usually printed on one side. Broadsides in the seventeenth and eighteenth centuries often filled the same niche in society that the front page or two of "scandal sheet" newspapers do today.

celestial equator   An imaginary circle in the sky formed by projecting the equator of Earth onto the sky. The celestial equator is ninety degrees from the celestial poles.

celestial poles   Imaginary points in the sky formed by projecting the Earth's north and south poles onto the starfield. The celestial equator lies midway between (ninety degrees from) the celestial poles.

coma (of a comet)   The more or less spherical cloud of gas and dust surrounding the nucleus of a comet. The material in the coma is dislodged from the solid nucleus by solar energy when the comet moves close to the Sun. The coma is the apparent head of the comet.

comet   An object in orbit around the Sun composed of dust embedded in frozen gases.

continuous spectrum   The spectrum provided by a glowing object that is solid, liquid, or dense gas. A continuous spectrum has all the colors from red to violet with no bright or dark lines.

corona (of the Sun)   The Sun's outer atmosphere. It begins about 5000 miles (8000 kilometers) above the Sun's apparent surface (photosphere) and can be considered to extend as far as the solar wind is still identifiable. The Earth may be viewed as traveling within the Sun's outer corona.

cosmic rays   Charged subatomic particles (mostly protons) moving through space at very high speeds, even approaching the speed of light. (Cosmic rays are particles, not light waves.)

dawn   Morning twilight; the period of partial light between full nighttime darkness and sunrise.

declination   A method of specifying the position of an object in the starfield. Declination in astronomy corresponds to latitude in geography: measuring positions north or south of the (celestial) equator in degrees.

differentiation   In an astronomical body, the separation of different chemicals from their original mixed state into different layers or regions.

direct revolution   The revolution of an object around the Sun in the same direction as the planets. (Also called prograde or posigrade revolution.)

direct rotation   The rotation (spin) of an object on its axis in the same direction as its orbital motion. (Also called prograde or posigrade rotation.)

dust tail (of a comet)   The tail portion of a comet composed of tiny solid particles. These particles are pushed out of the coma away from the Sun by the pressure of sunlight. The dust tail shines by reflecting sunlight and hence is yellowish-white. The dust tail is often noticeably curved.

ecliptic   The plane of the Earth's orbit around the Sun, hence also the apparent path of the Sun through the starfield (constellations of the zodiac) in the course of one year as the Earth revolves around the Sun.

ellipse   A closed geometrical figure formed when a plane (flat surface) cuts through a cone at an angle but does not cut through the base of the cone. Planets, asteroids, and comets orbit the Sun in ellipses.

emission (bright-line) spectrum   The spectrum provided by a glowing rarefied gas. An emission spectrum shows a series of bright lines that identify the constituent gases and their temperature.

ephemeris   A listing that gives a celestial object's positions for various times.

fluorescence   The process in which a gas absorbs and then emits light.

gas tail (of a comet)   *See* ion tail (of a comet).

geocentric   Centered on the Earth. Geocentric distances are measured from (the center of) the Earth. People used to think our solar system was geocentric, that is, the planets and Sun all revolved around the Earth.

giant planets   Jupiter, Saturn, Uranus, and Neptune.

gravitation   The attraction of matter for other matter.

head (of a comet)   The coma and nucleus of a comet.

heliocentric   Centered on the Sun. Heliocentric distances are measured from (the center of) the Sun. Our solar system is heliocentric, that is, the planets revolve around our Sun.

hyperbola   An open geometrical curve formed when a plane cuts a cone at an angle to the base greater than the slope of its sides. Because no comet inbound to the Sun has exhibited a definite hyperbolic orbit, astronomers believe that all comets are part of our solar system and not strays from elsewhere in the universe.

ion   An atom that has become electrically charged because of the loss or gain of one or more electrons.

ion tail (of a comet)   The tail portion of a comet composed of ionized (positively charged) gases. The ion (or gas) tail is formed by the interaction of particles in the solar wind with gases in the comet's coma. The ion tail shines by fluorescence and is usually straight and bluish-white.

Kepler's laws   The three laws that describe planetary (and cometary) motion and the motion of any celestial body around another. The laws (expressed for planets) are, in very general terms, as follows: (1) Planets travel around the Sun in ellipses; (2) When a planet is near the Sun, it travels faster than when it is far from the Sun; (3) The larger a planet's orbit around the Sun, the slower the planet travels and the longer its revolution takes.

long-period comet   A comet with an orbital period greater than 200 years.

magnitude   A measure of brightness in astronomy. The lower the magnitude number, the brighter the object. A star with apparent magnitude +1.0 is approximately 2½ times brighter than a star with apparent magnitude +2.0.

meteor   The luminous streak in the sky when a meteoroid burns up in the atmosphere (a "shooting star").

meteorite   The portion of a meteoroid that survives its collision with a planet or moon.

meteoroid   A small rocky object in space.

meteor shower   A display of many meteors caused by the Earth colliding with a stream of comet debris.

meteor storm   An exceptionally heavy meteor shower caused by the Earth colliding with a stream of comet debris.

minor planet   A planetlike body in the solar system too small to be classified as a planet. Most minor planets (or asteroids) orbit the Sun between Mars and Jupiter.

nebula   A celestial cloud of gas and dust.

node   A point along the orbit of an object where it crosses a reference plane. In the case of a comet, its "ascending node" is where it crosses the plane of the Earth's orbit (ecliptic) going north, and its "descending node" is where it crosses the plane of the Earth's orbit going south.

nongravitational force (operating on comets)   Vaporization from the comet's nucleus that provides a tiny rocketlike thrust that slightly affects a comet's orbital motion.

nucleus (of a comet)   The central mass of a comet, composed of frozen gases (mostly water) and dust-sized particles of rock.

Oort Cloud   A spherical swarm of trillions of comets surrounding our solar system. These comets orbit the Sun mostly at distances between 20,000 and 100,000 astronomical units.

parabola   An open geometrical curve formed when a plane cuts a cone parallel to the slope of the cone. Long-period comets have such extended orbits that their paths close to the Sun resemble parabolas.

parallax   A distance measure in astronomy determined by observing a nearby object from different positions to measure its displacement against a background of distant objects.

perihelion   The point in an object's orbit around the Sun at which it is closest to the Sun. (At this point, it is traveling at its fastest speed.)

periodic comet   An old name for a short-period comet (a comet that revolves around the Sun in less than 200 years).

photometry   The measurement of light intensities.

planet   An object that orbits a star and has too little mass for its gravity to create a nuclear reaction at its core. The fundamental difference between a star and a planet is mass. Comets are essentially tiny icy planets.

planetoid   Another name for a minor planet (or asteroid).

plasma   In physics, a gas made up of positive ions (charged particles) and electrons rather than neutral atoms or molecules. To keep the atoms at least partially ionized, the gas must be very hot.

plasma tail (of a comet)   *See* ion tail (of a comet).

polarimetry   The measurement of the amount that light is polarized. Polarization is a condition in light whereby the transverse vibrations of the rays assume different forms in different planes.

quasar   A very distant astronomical object that is pouring out vast quantities of energy. It is thought that quasars are the nuclei of young galaxies. The name "quasar" is an abbreviation of a quasar's telescopic appearance as a quasi-stellar object or a quasi-stellar radio source.

retrograde revolution   The revolution of an object around the Sun in a direction opposite to the planets.

retrograde rotation   The rotation (spin) of an object on its axis opposite to its orbital motion.

revolution   The motion of one object around another. The Earth revolves around the Sun in one year.

right ascension   A method of specifying the position of an object in the starfield. Right ascension in astronomy corresponds to longitude in geography: measuring positions east or west along the (celestial) equator. Right ascension measures eastward from the vernal equinox, usually in hours (24 hours = 360°; 1 hour = 15°).

rotation   The spin of a body on its axis. The Earth rotates once every twenty-four hours.

shooting star   Colloquial name for a meteor.

short-period comet   A comet with an orbital period less than 200 years.

solar wind   Fast-moving charged subatomic particles flowing outward from the Sun.

spectrogram   A photograph of a spectrum.

spectrograph   An instrument for photographing a spectrum.

spectroscope   An instrument for viewing spectra.

spectroscopy   The study of spectra.

spectrum   The array of colors or wavelengths obtained when light from an object is dispersed as it passes through a prism or diffraction grating. By examining a spectrum, scientists can tell the temperature, chemical composition, radial velocity, spin, and many other things about a light source and the material that lies between the source and the observer.

star   An object with enough mass so that gravity pressing on its center can (or once did) cause a nuclear reaction at its core. The fundamental difference between a star and a planet is mass.

sublimation   In chemistry, the passage directly from a solid to a gas without passing through a liquid state. (Frozen carbon dioxide [dry ice] sublimes at room temperature. Under the extremely low pressure conditions of space, water sublimes also.)

tail (of a comet)   Gases and solid particles from a comet's coma forced outward, away from the Sun, by the pressure of sunlight (dust tail) and the solar wind (ion tail).

transit   The passage of a small body (such as a comet) across the face of a large body (such as the Sun). (Transit has other meanings in astronomy as well.)

twilight   The period of partial light between full nighttime darkness and sunrise and between sunset and full nighttime darkness. Thus, twilight occurs in both the morning and evening.

vernal equinox   The point on the celestial sphere where the apparent motion of the Sun along the ecliptic crosses the celestial equator from south to north.

volatile   Vaporizes readily at a relatively low temperature.

zenith   The point in the sky that is directly overhead if you are standing vertically. (The zenith will be different for observers in different locations.)

zodiacal light   Sunlight reflected off tiny dust particles (primarily from comets) that lie mostly near the plane of the solar system (the zodiac). The reflection from the particles is most pronounced near the Sun, so zodiacal light is usually seen either after sunset or before sunrise. It is very faint and therefore can only be seen under dark sky conditions.

SUGGESTIONS
FOR FURTHER READING

Guides to Books and Articles

Freitag, Ruth S. *Halley's Comet: A Bibliography.* Washington, D.C.: Library of Congress, 1984.

Freitag, Ruth S. *Halley's Comet: A Selected List of References.* Washington, D.C.: Library of Congress, 1982 (regularly updated).

Books: Technical and Semitechnical

Brandt, John C., and Chapman, Robert D. *Introduction to Comets.* Cambridge, England: Cambridge University Press, 1981.

Hellman, C. Doris. *The Comet of 1577: Its Place in the History of Astronomy.* New York: AMS Press, 1971. (Chronicles evolving ideas about comets.)

Ponnamperuma, Cyril, Editor. *Comets and the Origin of Life.* Dordrecht, Holland: D. Reidel, 1980.

Wilkening, Laurel L., Editor. *Comets.* Tucson: University of Arizona Press, 1982.

Books for the General Reader

Calder, Nigel. *The Comet Is Coming!* New York: Viking Press, 1981.

Chapman, Robert D., and Brandt, John C. *The Comet Book.* Boston: Jones and Bartlett, 1984.

Denny, Norman, and Filmer–Sankey, Josephine. *The Bayeux Tapestry.* London: Collins, 1966.

Ronan, Colin A. *Edmond Halley: Genius in Eclipse.* Garden City, New York: Doubleday, 1969.

Articles: Technical and Semitechnical

A'Hearn, Michael F. "Chemistry of Comets." *Chemical & Engineering News,* Vol. 62, No. 22 (May 28, 1984), pp. 32–49.

Edburg, Stephen J. "International Halley Watch Amateur Observers' Manual for Scientific Comet Studies." Cambridge, Massachusetts: Sky Publishing, 1984.

Newburn, Ray L., Jr., and Yeomans, Donald K. "Halley's Comet." *Annual Review of Earth and Planetary Sciences,* Vol. 10, 1982, pp. 297–326.

Russell, Dale A. "The Mass Extinctions of the Late Mesozoic." *Scientific American,* January 1982, pp. 58–65.

Wetherill, George W. "Apollo Objects." *Scientific American,* Vol. 240, March 1979, pp. 54–65.

Whipple, Fred. "The Spin of Comets." *Scientific American,* Vol. 242, March 1980, pp. 124–34.

Whipple, Fred. "The Nature of Comets." *Scientific American,* Vol. 230, February 1974, pp. 49–57.

Articles for the General Reader

Alter, Dinsmore. "Comets and People." *Griffith Observer,* Vol. 20, July 1956, pp. 74–82.

Bortle, John E., and Morris, Charles S. "Brighter Prospects for Halley's Comet." *Sky and Telescope,* Vol. 67, January 1984, pp. 9–12.

Gould, Stephen J. "The Cosmic Dance of Siva." *Natural History,* Vol. 93, August 1984, pp. 14–19. (This article and the following one are valuable reviews of the mass extinction cycles controversy.)

Gould, Stephen J. "The Ediacaran Experiment." *Natural History,* Vol. 93, February 1984, pp. 14–23.

Klein, Jerry. "When Halley's Comet Bemused the World." *New York Times Magazine,* May 8, 1960, p. 45+.

Kerr, Richard A. "Periodic Impacts and Extinction Reported." *Science,* Vol. 223, March 23, 1984, pp. 1277–79.

Maran, Stephen P. "Where Do Comets Come From?" *Natural History,* Vol. 91, May 1982, pp. 80–83.

Olsen, Roberta J. M. "Giotto's Portrait of Halley's Comet." *Scientific American,* Vol. 240, May 1979, pp. 160–70.

Oppenheimer, Michael, and Haimson, Leonie. "The Comet Syndrome." *Natural History,* Vol. 89, December 1980, pp. 54–61.

Simon, Cheryl. "Death Star." *Science News,* Vol. 125, No. 16, April 21, 1984, pp. 250–52. (Good review of mass extinction cycles controversy.)

Stone, Greg. "A Sort of Heavenly Polywog." *Yankee,* Vol. 37, December 1973, pp. 90–97.

Yeomans, Donald K. *The Comet Halley Handbook* (2nd ed.). Pasadena, California: NASA, 1983. (Available from Superintendent of Documents, U.S. Government Printing Office, Dept. 33, Washington, DC 20402. Contains vital information for the casual or serious observation of Halley's Comet. Charts and diagrams are exceptionally helpful.)

INDEX

A

Copyediting and indexing by Karen McCeney
Cover and book design by Pamela Lewis
Production by Hilary Kanter
Managing Editor: Janet S. Dodd
Typeset by Hot Type Ltd., Washington, D.C.
Printed and bound by Maple Press Co., York, Pa.